Solar Energy

Books in the **Contemporary World Issues** series address vital issues in today's society such as genetic engineering, pollution, and biodiversity. Written by professional writers, scholars, and nonacademic experts, these books are authoritative, clearly written, up-to-date, and objective. They provide a good starting point for research by high school and college students, scholars, and general readers as well as by legislators, businesspeople, activists, and others.

Each book, carefully organized and easy to use, contains an overview of the subject, a detailed chronology, biographical sketches, facts and data and/or documents and other primary source material, a forum of authoritative perspective essays, annotated lists of print and nonprint resources, and an index.

Readers of books in the Contemporary World Issues series will find the information they need to have a better understanding of the social, political, environmental, and economic issues facing the world today.

Solar Energy

A REFERENCE HANDBOOK

David E. Newton

ABC-CLIO™

An Imprint of ABC-CLIO, LLC

Santa Barbara, California • Denver, Colorado

Copyright © 2015 by ABC-CLIO, LLC

Library of Congress Cataloging-in-Publication Data

Newton, David E.
    Solar energy : a reference handbook / David E. Newton.
        pages cm
    Includes bibliographical references and index.
    ISBN 978-1-61069-695-1 (acid-free paper) —
ISBN 978-1-61069-696-8 (ebook)   1. Solar
energy—Handbooks, manuals, etc.   I. Title.
    TJ810.5.N49 2015
    333.792'3—dc23   2015024828

ISBN: 978-1-61069-695-1
EISBN: 978-1-61069-696-8

19   18   17   16   15   1   2   3   4   5

This book is also available on the World Wide Web as an eBook.
Visit www.abc-clio.com for details.

ABC-CLIO, LLC
130 Cremona Drive, P.O. Box 1911
Santa Barbara, California 93116-1911

This book is printed on acid-free paper ∞

Manufactured in the United States of America

5   DATA AND DOCUMENTS, 217

The world is living in an Age of Fossil Fuels. Humans today depend very heavily on coal, oil, and natural gas to provide the energy needed for a host of purposes, from the heating of homes, office buildings, and other structures to a seemingly unending variety of industrial processes to nearly every form of transportation from automobiles and trucks to airplanes and ships. The Age of Fossil Fuels began with the rise of the Industrial Revolution in the late eighteenth century, and it is likely to continue as long as abundant supplies of coal, oil, and natural gas are available at reasonable cost. Some experts believe that period will come to an end relatively soon, perhaps within our lifetimes, while others say that the Age of Fossil Fuels may well continue for centuries. The one thing that nearly all experts agree with, however, is that the Age of Fossil Fuels cannot last forever. They agree that the supplies of fossil fuels are finite and that at some point in human history, other sources of energy must be found and developed.

Concerns about the finite (nonrenewable) nature of fossil fuels are hardly new. An article in the September 10, 1868, issue of *The New York Times*, for example, described what can only be described as a sense of urgency in Great Britain about the likelihood of the nation running out of coal in the near future, thus endangering the progress that had come about as a result of the Industrial Revolution. One response to this growing sense of panic among politicians, industrialists, engineers, and the general populace was a suggestion by the engineer John Ericsson (who is better known for his design of the Civil War ship *Monitor*) that scientists begin to study the use of solar energy as an alternative energy source. "The field awaiting the application of the solar engine is almost beyond computation," Ericsson once wrote, "while the source of its power is boundless. Who can foresee what influence an inexhaustible power will exercise on civilization?"

Interest in solar energy in the United States blossomed in the mid-1970s, when the Organization of Petroleum Exporting Countries (OPEC) declared an embargo on oil shipments to countries that had supported Israel in the Arab–Israeli war of 1973. Faced with a rapid and significant increase in the cost of oil, Americans and citizens of other nations affected by the embargo began to think about possible alternatives to oil as a source of energy. For many experts in the field, the answer to this challenge was solar energy. And during the 1970s, the U.S. Congress passed a number of bills encouraging and supporting the growth of solar energy facilities.

With the end of the OPEC embargo and the installation of a Republican presidential administration with greater interest in fossil fuels, interest in solar technology withered, not to arise again until early in the twenty-first century. At this point, concerns not only about the availability of fossil fuel supplies but also about the environmental effects of burning coal, oil, and natural gas once again forced solar energy to the forefront of national policy making. Significant improvements in the efficiency of solar photovoltaic (PV) cells and concentrating solar energy systems also contributed to renewed enthusiasm for the potential of solar energy as a factor in the nation's energy equation in the near and more distant future.

The acceptance of solar energy by policy makers and the general public has not been, however, unanimous and always enthusiastic. A number of questions have been raised about possible disadvantages in the expansion of solar power facilities. For example, most solar power plants require very large areas of land, which can impact agriculture, dairying, recreation, minerals extraction, and other essential activities. Also, the manufacture of solar PV cells produces a number of by products that may pose risks for the natural environment and human health. The question that continues to arise, then, involves the relative risks and benefits offered by solar energy as a major component of the nation's energy equation.

This book is designed to provide an overview of the field of solar energy, beginning with a history of human understanding and use of the technology from the earliest days of human civilization. Chapter 1 follows this history through the first stages of modern research on photovoltaic cells, active and passive solar heating, concentrating solar power systems, solar thermal heat, solar thermal electricity, and other types of solar technology. The chapter reviews the social, economic, political, and other issues related to this development of solar energy to the present day. Chapter 2, then, focuses on some of the major problems and issues associated with the development and use of solar energy, such as land use and human health issues, as well as questions about the economic viability of solar energy as a large-scale source of energy in the United States and other parts of the world.

Remaining chapters of the book provides resources for readers who want to continue their own research on solar energy. Chapter 3, for example, provides a variety of essays on very specific aspects of solar energy, including pro or con arguments with regard to the technology, information on new applications of the technology, and economic and social implications of the use of solar energy. Chapter 4 contains a list of important individuals and organizations in the field of solar energy, with descriptions of their contributions to and role in the development of solar energy. Chapter 5 provides excerpts from laws, court cases, reports, and other documents of relevance to solar energy, as well as tables of data about the production, consumption, and other aspects of the technology. Chapter 6 is an extended annotated bibliography. Chapter 7 is a chronology of important events in the history of solar energy, and the glossary defines important terms in the field.

Solar Energy

"But to truly transform our economy, protect our security, and save our planet from the ravages of climate change, we need to ultimately make clean, renewable energy the profitable kind of energy."

(President Barack Obama, 2009)

"Clearly, we need more incentives to quickly increase the use of wind and solar power."

(Former senator Hillary Rodham Clinton, 2005)

"The future is green energy, sustainability, renewable energy."

(Former California governor Arnold Schwarzenegger: Green Quest Goes On 2012)

Renewable energy is an issue of significant concern in the world today. One reason for that concern is the expectation that the fossil fuels on which humans have relied so heavily in the past two centuries—coal, oil, and natural gas—are being depleted and may not last much longer. The world will soon have to find other sources of energy to replace those diminishing supplies of fossil fuels. Also, the use of fossil fuels for heating, transportation, industrial operations, and other purposes has been releasing unprecedented amounts of carbon dioxide and other

---

Cliff Palace, an Anasazi settlement at Mesa Verde National Park. This large structure, highlighting the southern-facing cliff villages accessible to its residents only, by climbing up the rocks, housed about 100 people. The cliff villages serve as an excellent example of early passive solar design. (Corel)

greenhouse gases to the atmosphere to the point that the planet's climate has begun to change in dramatic and worrisome ways. Again, renewable sources of energy appear to be the one major way of dealing with this serious environmental problem.

And so, national governments, research institutions, and industrial corporations have begun to search for viable alternative and renewable sources of energy, such as wind, water, geothermal, and solar energy, that can be developed for wider human use.

But renewable energy is hardly a new concept. For much of early human history, men and women relied on the power of wind and running water to carry out the many forms of work with which they were faced every day. Wind power was used to propel boats across the water and turn windmills, while the energy provided by streams and rivers raised water for irrigation and turned water wheels to grind grain. Renewable energy for humans in the pre-Christian era was not some exotic topic of little practical value; it was a part of the core of energy resources on which human life depended.

Solar Energy: The Early History

It is difficult to imagine that there was ever a time that humans did not understand and appreciate the contribution that the Sun makes to the survival of the race. Humans must have realized that solar energy was responsible for the warmth that allowed them to survive through warm months of the year, and that the promise that those months would return again at some time in the future kept alive early human societies. It must also have been obvious that sunlight was responsible for the growth of plants, ultimately the source of virtually all life on Earth.

Sun Worship

So it is hardly surprising that Sun worship arose during the earliest periods of human civilization and that it remained for many centuries one of the most important—if not *the* most important—of all early human religions. As one scholar has written, "The pre-eminence of the Sun, as the fountain-head of life and man's well-being, must have rendered it at a date almost contemporaneous with the birth of the race, the chief object of man's worship" (Olcott 2008, 141).

Archaeologists have now determined that Sun gods and goddesses were integral parts of almost every early civilization. Some cultures adopted multiple Sun gods and goddesses, generally with somewhat different responsibilities. Ancient Egypt, for example, honored Ra (or Re) the supreme god of the country; Aten, the god of the Sun's visible disc; Horus, whose right eye was thought to be the Sun and his left eye, the Moon; Nefertem, who represented Sunrise; Atum and Khnum, the gods of Sunset; and Sopdu, the god responsible for the extreme heat typical of Egyptian summers (Vendel 2015). Hinduism also had a panoply of gods and goddesses associated with the Sun, including the chief Sun god Surya; Varuna, who causes the Sun to travel across the sky; Savitr, the god of the Sun before it rises; Rta, who is responsible for maintaining the orderly rising and setting of the Sun (among other natural phenomena); Agni, the god of fire (so often associated with the Sun itself); and Saranyu, the goddess of the dawn (Jayaram 2015). Similar lists of solar deities exist for many African, Mesoamerican, and East Asian cultures. Sun gods and goddesses were by no means restricted to tropical regions, as is attested by the range of deities found, for example, in Celtic, Norse, and Slavic mythology. (For an extensive list of Sun gods and other deities, see "Names of Gods and Goddesses" 2008.)

The designation of certain solar deities was not an abstract theological concept for early cultures but a fundamental and defining feature of such societies. Adoration of the gods, maintenance of their status in a culture, ensuring that their earthly needs were always being met, and retaining all the other vestiges of which any god or goddess is worthy almost inevitably required the provision of a large collection of priests and other supernumeraries to carry out all the routine and special events conducted in honor of the deity or deities. Indeed, in many cultures the priesthood was virtually indistinguishable from the secular government itself. In ancient Egypt, for example, the pharaoh was both king and chief priest.

Solar religions also required a significant investment in material structures, such as the construction of temples and other buildings in which religious ceremonies could be conducted and where the sometimes very large priesthood could be housed. One of the interesting features in so many of these buildings was the use of designs that would ensure that certain solar events would always occur at the same time and in the same way. One of the oldest and best known of these buildings is a structure located in County Meath, Ireland, thought to have been built in about 3200 BCE. The structure is oriented in such a way that sunlight enters the building through a specially built hole and reaches the innermost chamber of the structure only one day a year, on the winter solstice. Archaeologists do not know what the function of the building was, but the essential role of sunlight in its activities is certainly clear (O'Kelly 1998). Many other such structures are known, including the megaliths at Stonehenge, England; the El Castillo and Uxmal pyramids in Mexico; the Maeshowe tomb in Scotland; and the Great Kiva at Chaco Canyon, New Mexico (for more detail on such structures, see Taylor 2012).

As critical a role as it may have had in religion, in architecture, and in other fields of ancient civilizations, the power of the Sun played essentially no part in meeting the energy needs of these societies. The Babylonians, Assyrians, Persians, Chinese, Japanese, Egyptians, and other early cultures *did* rely on some forms of renewable energy to meet their everyday working needs but not on solar energy. For example, these cultures often had highly developed systems for capturing the wind and using it to propel their ships at sea and to turn primitive windmills. And they had sophisticated systems for raising and moving water needed for often-elaborate irrigation system. But they knew relatively little about ways to capture and concentrate solar energy so that it could be used to perform work.

Passive Solar Heating Systems

The one major exception involved the use of passive solar heating systems to heat buildings. The term *passive solar heating system* refers to a system that makes use of no mechanical devices or moving parts. The simplest example of a passive solar heating system in today's world might be the family car sitting at rest in the driveway. If the Sun is hot, solar radiation passes through the windows of the car, heating up its interior. No one has to do anything to make this system work; it converts solar energy into thermal energy (heat) simply by passing through the car windows and warming the interior.

Archaeologists now know that many early civilizations were familiar with the principles of passive solar heating and used these principles to build their homes, commercial buildings, and even whole cities. The earliest reliable records that exist for the use of passive solar heating for homes come from about the fifth century BCE, when a number of commentators describe a common practice whereby families built their homes with an open space facing toward the south. The opening was covered so that the Sun's rays could enter the house in the winter, when it was low in the sky, but was reflected by the roof in the summer, when the Sun was higher. This design made it possible to use solar radiation to keep the house warm in winter and cool in summer. As Socrates is said to have remarked about this design, "In houses that look toward the south, the sun penetrates the portico in winter, while in summer the path of the sun is right over our heads and above the roof so that there is shade" (Butti and Perlin 1980, 5; this reference contains an illustration of an ancient Greek passive solar structure). Houses with this type of design today are sometimes known as Socrates megaron houses, named after the Greek philosopher, and a rectangular space usually found in palace complexes.

Passive solar heating was an important technology for Greek architects not only because sunlight was free and designs were relatively simple. In addition, the Greeks were at the point of running out of arguably their most important single source of energy for heating—wood—by the fifth century BCE. The growth of cities and an expanding population had largely denuded the country of its native forests, and wood for heating was rapidly becoming a rare and expensive commodity (with perhaps a lesson here for modern societies) (Botkin 2010, 12).

In any case, passive solar heating was a concept of more than casual importance to the Greeks. Indeed, they went so far as to design whole cities based on the principle. Probably the best known of these cities was Olynthus, a city of about 2,500 inhabitants. Although it had existed since the seventh century BCE, it was resettled and expanded in 432 BCE by King Perdiccas II of Macedon as a reward for the city's inhabitants' revolt against Athens. The new portion of the city was built to be entirely solar-friendly, with streets laid out in a north-south direction to allow all houses to be aligned with the Sun's rays, as described earlier. (Olynthus has been well studied, and a number of helpful references are available, including Cahill 2008; Perlin 1986; and Robinson and Graham 1938. For a city design and layout of Olynthus and other solar cities, see Butti and Perlin 1980, 6, 7, 8).

Neither was Olynthus the only city specifically designed to take advantage of passive solar energy for the heating of buildings. Other cities with similar histories include Priene, Delos, and parts of Athens itself. A prominent historian of the period, John Perlin, has written that "[p]assive houses appear to have been so common in ancient Greece that [historian and student of Socrates] Xenophon uses them to demonstrate the Socratic notion of beauty" (Perlin 1986, 330).

Passive solar heating was one of the many technologies passed on from Greek to Roman civilizations around the beginning of the new millennium. Perhaps even more than the Greeks, the Romans had begun also to experience a loss of wood resources, which they used for a host of purposes, such as the construction of homes and other buildings, for building ships, and to heat their homes and their beloved public and private bathing areas. According to available records, operating a *hypocaust*, or hot-air heating system, in a large private villa could require the combustion of nearly 300 pounds of wood *per hour*, or more than two cords *per day* (Butti and Perlin 2008, 15).

As with the Greeks, the Romans responded to this impending disaster by turning to solar energy as a substitute for wood burning. In fact, it appears that passive solar homes became at least as common in the early Roman Empire as they had been in classical Greece. As the scholar Marcus Terentius Varro wrote in the first century BCE, "What men of our day aim at is to have their winter rooms face the falling sun [the southwest] as a way of keeping their homes warm during the coldest part of the year" (Perlin 2005).

As to be expected, the Romans made a number of improvements on the use of solar energy, as they did with other technologies "borrowed" from the Greeks. Perhaps the most important of these was the use of mica as a transparent material for enclosing spaces heated by the sun. (Plate glass was not available since it was not invented until the seventeenth century.) The precise details of the occasion on which mica was first used as a transparent window for passive solar designs are not known. According to some of the best evidence available, however, such windows may have been used in the solar homes built for the wealthy author and politician Pliny the Younger in the first century CE. Pliny called rooms covered with mica windows *heliocamini* (singular: *heliocaminus*), or "solar furnaces," because the heat produced was much greater than in open passive solar spaces (Butti and Perlin 2008, 19). Before long, any Roman wishing to build a passive solar house—and who could afford the extra expense—called for the use of mica windows in the structure.

A second major development in the use of passive solar systems came in the field of law. As solar structures became more popular throughout the empire, some practical problems began to arise that interfered with the success of the technology. Consider the case in which the Gratius family, for example, decided to build a passive solar home in the outskirts of Rome. And then imagine that a second family, the Laberii, later decided to build their own home on land next to the Gratii and designed a structure three stories taller than the Gratii home. Chances were good that the Laberii home would block out the Sun's rays from the Gratii home, for at least part of the day. The Laberii had, probably inadvertently, made it impossible for the Gratii family to heat their home adequately.

As passive solar heating became more popular, disputes such as the imaginary one between the Gratii and Laberii more and more often reached the courts. Finally, in the second century CE, a famous and respected jurist, Ulpian (Gnaeus Domitius Annius Ulpianus) ruled that a builder could not interfere with the solar radiation falling on a neighbor's solar facility. As more and more decisions of this kind were announced, a formal Roman policy on solar access eventually became part of the Justinian code, promulgated between 529 and 534 CE. It became illegal for anyone to disrupt a person's access to solar radiation, either through the construction of a building or by allowing trees to grow too tall (Jordan and Perlin 1979–1980; for the relevant part of this code, see Chapter 5, Data and Documents).

So-called solar access laws, such as those contained in the Justinian code, survived in many places for more than a thousand years. They have slowly been replaced in modern society, however, by more complex and more sophisticated laws and regulations about access to solar radiation for solar heating systems, as will be discussed later in this book.

Other Uses of Solar Energy

The ancients also found a number of other uses for solar energy, some of which depended on devices for concentrating the Sun's rays, as with a concave mirror or lens. Archaeologists have uncovered evidence for the use of glass lenses and metal sheets shaped to focus the Sun's rays, possibly for the purpose of setting objects on fire on lighting a flame. One of the oldest of these devices is the so-called Nimrud or Layard Lens, currently held by the British Museum. The lens consists of an oval piece of rock crystal 4.2 cm by 3.45 cm and 0.25 cm thick. The crystal has been polished to produce a flat surface on one side and a convex surface on the other side, similar to a very simple magnifying glass that would be available today. The object is dated to somewhere between 750 and 710 BCE. The discoverer of the lens claimed that it is "the earliest example of a magnifying and burning-glass" (Layard 1853, 197–198; for an image of the lens, see "The Nimrud Lens/The Layard Lens" 2015; for further information about the use of magnifying lenses in the ancient world see Hanna 2010).

A reminder of the way that lenses were used by the ancient Greeks is a frequently quoted passage from "The Clouds," a play written by the Greek playwright Aristophanes in 419 BCE. In the play, the main character, a man named Strepsiades, describes to the philosopher Socrates a scheme he has developed to escape a warrant that has been issued for his arrest:

STREPSIADES I have found a very clever way to annul that conviction; you will admit that much yourself.

SOCRATES What is it?

STREPSIADES Have you ever seen a beautiful, transparent stone at the druggists', with which you may kindle fire?

SOCRATES You mean a crystal lens.

STREPSIADES That's right. Well, now if I placed myself with this stone in the sun and a long way off from the clerk, while he was writing out the conviction, I could make all the wax, upon which the words were written, melt.

SOCRATES Well thought out, by the Graces! (Aristophanes 419 BCE)

Of all the stories from the ancient world told about the use of mirrors, lenses, and other reflective materials for the capture and utilization of solar energy, none has probably been as widely quoted as an invention attributed to the Greek mathematician, philosopher, and inventor Archimedes, who died in 212 BCE. According to this tale, Archimedes had been asked to use his skill as an inventor to help protect his home country, Syracuse, against the attacking Romans in the Punic Wars that raged for almost a century, from 264 BCE to 146 BCE. Among the many devices that Archimedes is said to have invented was a collection of "burning mirrors." The device supposedly consisted of a number of mirrors oriented in such a way as to capture sunlight and focus it on Roman ships in the harbor of Syracuse. The tale concludes with confirmation of the fact that the concentration of solar radiation produced by the mirror system was sufficient to set fire to the Roman ships and save the city of Syracuse from invasion. (For one possible design of such a system and a description of its use, see Krystek 2011.)

(The debate as to whether such a system could ever have actually been built and implemented has raged on for the past two thousand years with experts on both sides of the question still arguing for or against the possibility of Archimedes's invention having been true. For a good review of that history, see Wik 2013, Chapter 1.)

A Solar Renaissance?

With the rise of the Roman Empire and the arrival of the new millennium, it might have seemed that an age of solar energy was about ready to dawn on the world. Although somewhat modest, the inventions of passive solar heating and the concentration of sunlight with mirrors and lenses would seem to have provided the basis for even more sophisticated and useful solar devices that could be put to work for humans. But such was not to be the case. As intellectual pursuits in Western Europe turned toward religious concerns and away from the necessities of everyday life, research in solar energy largely disappeared. In fact, it was not until the early eighteenth century that inventors and scientists once more began to think about the ways in which solar radiation could be captured, concentrated, and put to work in new kinds of devices and machines.

The one exception to that statement is a single invention announced in 1615 by French engineer and inventor Salomon de Caux (also, de Caus), a solar-powered water pump. The pump consisted of a metal container filled with water and air suspended in a wooden frame. When sunlight was focused on the container, the air inside expanded forcing out water in the form of a fountain. De Caux imagined a number of applications for his steam-powered machine, including one of the first motorcars, but none of them came to a realization, and his steam engine was regarded largely as an imaginative and interesting toy. (De Caux verbally described and drew images of his earliest steam engines with some possible applications in his historic book *Les Raisons des forces mouvantes* [*Reasons for Moving Forces*], published in 1615. *See* Caus 1615, for example, at http://cnum.cnam.fr/CGI/fpage. cgi?FDA1/49/100/162/0162/0162 and http://cnum.cnam.fr/ CGI/fpage.cgi?FDA1/51/100/162/0162/0162.)

More than a century and a half later, another breakthrough in the use of solar energy was announced in 1767 by French physicist Horace Benedict de Saussure, a device he called a "solar hot box," which we might call today a *solar cooker*. De Saussure's invention was made possible at least in part by the invention of plate glass, which had first been made commercially toward the end of the seventeenth century. He utilized the new material to build a stack of five boxes placed inside each other, each with walls made of plate glass. De Saussure found that when this device was placed in direct sunlight, each of the boxes was heated to a warmer temperature, from the outermost box to the innermost. The heat generated in the innermost box was, in fact, sufficient to boil water and cook foods, although the device itself and the challenge of keeping it oriented toward the Sun largely negated its practical use (Butti and Perlin 1980, Chapter 5).

De Saussure's hot boxes did not exactly revolutionize the practice of cooking, but they did not go completely unappreciated either. In their unmatched book on the history of solar energy, Ken Butti and John Perlin note that two later world-famous physicists picked up on de Saussure's ideas and built improved models of his hot boxes. In the 1830s, for example, British astronomer Sir John Herschel built and carried with him a hot box of his own design on excursion to South Africa, in which he regularly cooked his own meals. A half century later, American astrophysicist Samuel Pierpont Langley devised his own modified form of a hot box, which he also carried with him on his research expeditions (Butti and Perlin 1980, 57–59).

An Age of Solar Power?

Perhaps the preeminent proponent of capturing solar radiation for the operation of machines was a French high school mathematics teacher, Augustin Mouchot. Mouchot was living at a time when the Industrial Revolution was bringing humans previously unimagined and apparently unlimited improvements in their lives, from new forms of transportation to industrial operations designed to meet every conceivable need or want. But Mouchot looked further down the road and worried about a future in which industry and transportation would continue to demand ever and ever greater amounts of coal in order to operate. On one occasion, he is said to have observed that "[e]ventually industry will no longer find in Europe the resources to satisfy its prodigious expansion. . . . Coal will undoubtedly be used up. What will industry do then?" (This quotation is to be found in many sources, but no specific citation is apparently available. See, for example, Cleveland and Morris 2015, 666.)

Mouchot believed that the answer to this question was to "harvest the sun." He spent the better part of two decades, from 1860 to 1879, experimenting with a variety of ways of capturing sunlight, concentrating its rays, and using that radiation to boil water. The steam produced in these operations was used to drive a variety of mechanical devices, ranging from simple cooking equipment to desalination plants.

Probably Mouchot's greatest step forward was his melding of two solar technologies previously developed by other inventors. The first technology made use of de Saussure's concept of collecting solar radiation inside nested glass containers, each of which was hotter than its outer companion. The second technology made use of mirrors of various shapes that could focus a beam of sunlight onto a single spot, concentrating all of the solar energy on one specific point, the focus of the mirrors. When the focus of the mirror system was located inside a nested set of glass boxes, very high temperatures could be produced in a relatively short period of time. (For images of Mouchot's various solar engine designs, see Butti and Perlin 1980, Chapter 6.)

One of Mouchot's first inventions was the so-called *marmite solaire* ("solar cooking pot"), for which he received a patent in 1861. The device was simplicity itself, consisting of a parabolic metallic trough aligned with its axis parallel to a cylindrical glass container. The trough reflected and concentrated sunlight into the cylinder, where temperatures reached a few hundred degrees Celsius, sufficient to boil water, cook food, and melt some metals. (For a more detailed description of this and other Mouchot inventions, see Collins 2002; Simonin 1876.) Over the next two decades, Mouchot worked on a variety of refinements of this basic design, culminating in a series of massive engines capable of converting solar energy into steam for powering large machines with impressive efficiency. Along the way, he continued to receive support and encouragement from the French government in the latter's efforts to give at least some attention to the question of how to maintain its industrial system if and when its coal supplies began to run out.

Mouchot realized his first peak accomplishment in 1869, when he unveiled the largest solar engine ever built, with a copper boiler seven feet in diameter, capable of producing a head of steam of 45 pounds per square inch (see Collins 2002; for an image of the device, see "19th Century Solar Power Engines" 2012). It was in the same year that Mouchot published what is generally regarded as the first book on solar energy, *La Chaleur Solaire et les Applications Industrielles* (*Solar Heat and Industrial Applications*). Mouchot's momentous device disappeared when the Prussian army invaded Paris in 1871, but as soon as he was financially able to do so, he rebuilt a similar device in his hometown of Tours. The enormity of the machine seems to have been astounding, as recorded by an observer writing for the journal *Revue des Deux Mondes* (*Review of the Two Worlds*) in 1876. The reviewer noted that the steam pressure created by Mouchot's device was so great that it threatened to blow up the boiler itself (Simonin 1876, 555).

Mouchot, the French government, and many proponents of solar power saw a promising future for Mouchot's inventions. The author of the previous review expressed his own optimism about the future of solar power. He noted that

> it is especially in tropical countries that it is destined to find immediate employment, in driving the various kinds of machinery used in sugar and cotton plantations, in distilling impure water to make it fit for drinking, in crystallizing saline and saccharine solutions, in pumping water of irrigation, in manufacturing ice by means of the Carré machine, etc. In those countries fuel is scarce, firewood is not abundant, and coal, which has to be imported from a distance, often from the mines of England, commands an exorbitant price. Already in southern countries sea-salt is obtained purely by the action of solar heat. In Chili and in Mauritius, salt-marshes are divided into compartments, with walls and roof of glass, in order to promote evaporation; so in the famous nitre-beds of Iquique, on the coast of Peru, the salt might be crystallized by solar heat alone. (Simonin 1876, 556)

Mouchot's most remarkable achievement was probably a solar engine that he and his assistant, Abel Pifre, built for the Paris Exhibition of 1878. The device consisted of a 21-gallon boiler powered by solar radiation reflected from a 13-foot-diameter mirror. Perhaps the most startling feature of the machine to fair-goers was that it was used to make ice, an accomplishment that seems to have impressed the fair judges also, who awarded Mouchot a Gold Medal for the machine (Butti and Perlin 1980, 72; for an image of the machine, see "The 19th Century Solar Engines of Augustin Mouchot, Abel Pifre, and John Ericsson 2012).

By the time the Paris Exhibition had come to an end, economic conditions in France and the rest of the world had begun to change. The price of coal in England had begun to drop significantly, and the French government (among others) had decided that solar power was too uneconomical and too risky to deserve further investments. Discouraged by these changes, Mouchot decided to end his research on solar energy and return to his teaching job in Tours. That decision by no means meant an end to research on solar power, however. A number of other researchers in France and other parts of the world retained their optimism about the future of solar power and continued their research in the field. For example, Mouchot's assistant, Pifre, constructed a solar engine in the Tuileries Gardens in Paris with which he powered a printing press that produced 500 copies an hour of a newspaper he called *Le Journal du Soleil* (the *Sun Journal*). For an image of the machine, see "The 19th Century Solar Engines of Augustin Mouchot, Abel Pifre, and John Ericsson 2012."

Even with the falling prices of coal, the potential for solar power to take over the operation of machines from coal remained a driving force for individual inventors from a number of countries. In the United States, for example, one of the first persons to search for an economical way of harvesting solar power was Swedish-born inventor John Ericsson. Literature is replete with glowing comments about the potential of solar energy from Ericsson. In one article about solar power by one of his colleagues, for example, Ericsson is quoted as predicting that "[t]he field awaiting the application of the solar engine is almost beyond computation."—"Who can foresee what influence an inexhaustible power will exercise on civilization?" (Willsie 1909, 512). Like Mouchot and others, Ericsson was concerned about the almost-complete dependence of the Industrial Revolution on coal. In his four-hundred-page doctoral thesis for the Lund University, he wrote:

> Already Englishmen have estimated the near approach of the time when the supply of coal will end, although their mines, so to speak, have just opened. A couple of thousand years dropped in the ocean of time will completely exhaust the coal fields of Europe, unless, in the meantime, the heat of the sun be employed. (as quoted in Church 1890, vol. 2, 265)

Ericsson was a prodigious inventor with dozens of patents in a variety of fields. At first, he attacked the problem of solar power along much the same lines as those of his French predecessors, although he was, at the same time, largely dismissive of their accomplishments in the field, calling their inventions "mere toys" (Collins 2002). Eventually, he also decided to try a fundamentally different approach, essentially going back to de Saussure's original concept of a hot box. Instead of focusing the Sun's rays on a container of water, as Mouchot and his colleagues had done, Ericsson decided to just use air. The goal was to use solar radiation to heat up a container of air, which expanded as it became warmer, performing the same function as did steam in the Mouchot machines. (For an image of the Ericsson hot-air device, see Lienhard 2013, Figure 12.) In a letter he wrote in 1873 describing his work, he noted:

> I have this day seen a machine actuated by solar heat applied directly to atmospheric air. In less than two

minutes after turning the reflector toward the sun the engine was in operation, no adjustment whatever being called for. In five minutes maximum speed was attained, the number of turns being by far too great to admit of being counted. . . . the above solar engine is operated without valves, and is therefore absolutely reliable. As a working model, I claim that it has never been equalled. (Church 1890, vol. 2, 268)

Ericsson had his sights trained in particular on farmers in the arid western states as potential buyers for his new solar engine, especially since coal was difficult to obtain and expensive to buy in the region. Unfortunately for him, however, in spite of considerable interest from Westerners, his device was not yet sufficiently developed to put into production by the time he died in 1888. Shortly before his death, he had described to a friend his situation:

It will be proper to mention that the successful trial of the sun-motor attracted the special attention of landowners on the Pacific coast, then in search of power for actuating the machinery needed for irrigating the sunburnt lands. But the mechanical details of connected with the concentration at a single point of the power developed by a series of reflectors, was not perfected at the time. . . . Consequently, no contracts for building sunmotors could then be entered into. . . . (Church 1890, vol. 2, 275)

The full story of the development of solar power devices at the end of the nineteenth and beginning of the twentieth centuries requires far more space than this book can provide. One of the interesting threads in that story is the gradual evolution of design as each succeeding inventor made his (and it was apparently always men involved) machine a bit more efficient than that of his predecessors, often while depending on the basic design of that predecessor. For example, French inventor Charles Tellier proposed a new approach to the design of solar engines in the late 1800s, when he suggested the use of low-temperature liquids, rather than steam, as the working fluid in the engines. Prior to Tellier's time, virtually all solar engines used solar energy to convert water to steam, which was then used to operate a machine. Tellier suggested using liquids that vaporize at lower temperatures, such as ammonia, which boils at –33°C (–28°F), and sulfur dioxide, which boils at –10°C (14°F). In Tellier's solar engine, sunlight was shined on a container filled with liquid ammonia or sulfur dioxide, which then vaporized. The gas produced was then piped to the location at which it was to perform work. In the first of Tellier's inventions, the gas exerted pressure on water in a container, forcing the water out in the form of a jet (Smith 1995).

Tellier was also involved in a variety of other inventions and lost interest in his solar engine fairly soon. (He is also known as the "father of refrigeration" for his work in that field.) But other inventors took up his suggestion for low-boiling-point liquids as working liquids in solar engines. Many of them also followed another common thread of inventions for the period: the realization that the most promising region for the use of solar machines was likely to be the American West in general, and the Southwest in particular. An ongoing problem in this area was finding a source of power for the movement of water through irrigation systems, on which the region's newly burgeoning agricultural industry depended.

Two American inventors in particular who followed these two threads in their research on solar power were Henry E. Willsie and John Boyle, Jr., both from St. Louis. Willsie and Boyle adapted Tellier's design for a solar engine so that it would also solve one of the enduring problems of such devices, storage. Solar power can be collected only during the daytime, of course, so what can be done to store the energy collected during the day through the evening hours so that it will not just be lost during nonproductive hours? Willsie described the team's answer to that question in a now-classic paper published in the journal *Engineering News*, in May 1909, "Experiments in the Development of Power from the Sun's Heat" (Willsie 1909).

The system described by Willsie was fairly simple. Solar radiation is captured by a tank with a glass to and filled with water drawn from a well. The heated water is piped from the tank into a second holding tank that is well insulated to keep it from losing heat. That warm water is then circulated through a "boiler" filled with liquid sulfur dioxide, which is converted by the water to a gas. The gaseous sulfur dioxide then becomes the working fluid that operates some type of machine. When the sulfur dioxide does work and gives up heat, it changes back to a liquid, which is recirculated through the system. (The system is described in a contemporaneous article in *Scientific American* magazine; see "Engine Power from Solar Heat" 2013/1909.)

Willsie reported in his article that his device collected 2,300 Btu of thermal energy per day in the summer and 1,600 Btu in the winter, enough energy to produce a boiler pressure of 215 pounds per square inch and 20 horsepower of power. Willsie estimated that his plant would cost about $164 per horsepower to operate, compared to $40 per horsepower for a conventional steam plant. This differential was an obvious reason that Willsie and Boyle experienced serious problems in developing and marketing their solar engines, which never achieved any significant commercial success. (For a detailed explanation of the operation of the Willsie–Boyle engine, see Abbott 1938, 207–209.)

Willsie and Boyle were by no means the only inventors to imagine significant commercial applications for their inventions. In the early 1890s, for example, Baltimore home-heating salesman Clarence Kemp invented a device that heated water by means of solar radiation and then kept that water warm overnight in a holding compartment. Kemp advertised the system at first to husbands who were left at home alone for a period of time and presumably were not able to deal with more conventional methods of heating water than were their wives. Kemp's Climax solar water heater was briefly a commercial success in some parts of the country, especially in the western states, where sunshine was abundant. At one point, it is said that a third of all the homes in Pasadena were heating their water with a Climax device by 1900. Kemp's business success was short-lived, however, and he was soon out of the solar home-heating business. (Butti and Perlin 1980, Chapter 10, which also contains diagrams of the Climax heater and associated devices.)

Another solar entrepreneur was Aubrey Eneas, a British citizen who had immigrated to the United States and settled in Boston. He became interested in solar power at an early age and began to work on devices that could be used to convert solar power to mechanical energy. In 1892 he founded the Solar Motor Company in Boston with the goal of making and selling devices for a variety of purposes, the most important of which was moving water through irrigation systems in the West. Although his invention did meet with some short-term success, it probably became most famous for the display of solar motors at an ostrich farm near Pasadena, California. Tourists were invited to visit the farm, not only to see the huge birds but also to observe in action the massive solar devices at work on the farm. Like essentially all of his predecessors and successors in the business, Eneas soon came to the realization that solar energy could not compete with conventional fossil-fueled engines and he went out of business (Butti and Perlin 1980, 81–89).

The fundamental challenge facing solar energy researchers in the United States and most other parts of the developed world was that the resource they were attempting to use—solar energy—while certainly abundant and free—could not be turned into useful energy at a price that was even close to being competitive with more conventional energy sources, such as coal, petroleum, and natural gas. As the twentieth century developed, more and more reserves of these fossil fuels were discovered and exploited, and the price for all three fuels continued to either decline or remain well within the range that consumers were willing to pay. By contrast, the price of renewable energies such as solar, wind, geothermal, and hydroelectric continued to remain at levels far beyond those that could compete with the fossil fuels. (As an illustration of this point, notice the growth in consumption of the conventional fossil fuels during the twentieth century as shown in History of Energy Consumption in the United States, 1775–2009, 2011.)

The Development of Photovoltaic Power

Parallel with research on finding ways to capture solar energy for the operation of various types of engines, a separate line of inquiry concerned an entirely different aspect of solar engine, namely, the ability to convert solar energy into electrical energy. Probably the first experiment in that line of research was carried out in 1839 by French physicist Alexandre-Edmond Becquerel, a member of a distinguished line of researchers that included his father, Antoine César Becquerel, professor of physics at the National Museum of Natural History in Paris, and his son, Henri Becquerel, discoverer of radioactivity. At the age of 19, Edmond had become interested in the study of voltaic cells, chemical systems in which potential differences between two different metals can be used to generate an electric current (the basic principle behind, for example, the operation of all batteries).

In 1839, Edmond Becquerel had designed a variation on this research in which he immersed two sheets of platinum metal in an acidic solution and allowed light of various types to fall on the platinum metal. He found that the presence of light induces the flow of electricity between the two metal plates, a phenomenon that was affected by the type of light used (e.g., sunlight versus blue or red light). That phenomenon is now generally known as the *photovoltaic effect* (from *photo-* for "light" and-*voltaic* for an electric current produced by a chemical reaction). (For a diagram of Becquerel's experiment, see "First Photovoltaic Devices" 2015.)

To be completely correct, one needs to distinguish between the *photovoltaic effect*, discovered by Becquerel, and the *photoelectric effect*, discovered in 1887 by German physicist Heinrich Hertz. The former term refers to the production of an *electric potential* or *current* produced when a piece of metal is exposed to light, while the latter term refers to the production of *electrons* by that process. Of course, a current is a flow of electrons, so the terms are similar. But in many instances, it is necessary or desirable to keep the two terms distinct from each other. For a more detailed discussion of this point, see "The Photoelectric and Photovoltaic Effects" 2015.

Becquerel was able to find one specific application for his discovery, the invention of a device known as an actinograph. The actinograph was used to measure the temperature of a heated body by determining the amount of light it produced. Otherwise, Becquerel's discovery can hardly be said to have revolutionized the science of solar studies. Indeed, it was about a half century before a discovery of any significance extended scientists' understanding of the photovoltaic effect. Then, in 1873, British engineer Willoughby Smith made an important discovery purely by accident. In working on the laying of a submarine transatlantic cable from England to the United States, Smith was looking for a metal with a high resistance to electrical flow for use in the system. He decided to use selenium, a material whose conductivity is roughly one billion times that of copper. His experiments with the material, however, revealed a surprising fact, namely, that the conductivity of selenium increases dramatically when it is exposed to light. Today, one would say that the material becomes *photoconductive*, capable of carrying an electrical current in the presence of light (see "Effect of Light on Selenium during the Passage of an Electric Current 1873"; Smith 1873).

Only three years later, English physicists William Grylls Adams and Richard Evans Day made an interesting discovery in pursuing Smith's findings about selenium. They connected a bar of selenium to two platinum wires at opposite ends of the bar inside a glass tube and shined light on the system. They discovered that the presence of light causes the flow of an electrical current through the system, the first time that the photovoltaic effect had been observed in a dry system (compared to the solution used by Becquerel). Adams and Day noted the primary conclusion of their research, namely:

> This experiment was repeated in various ways with light from different sources, the results clearly proving that by the action of light alone we could start and maintain an electrical current in the selenium. (Adams and Day 1876, 115)

Unlike Becquerel's discovery, the research of Adams and Day set off a flurry of experiments on the effect of light on selenium in various configurations. Probably the most productive of these experiments was a series conducted by American physicist and inventor Charles E. Fritts. In the early 1880s, Fritts constructed the first solar cell capable of converting sunlight directly into electrical current. The cell consisted of a thin sheet of selenium metal coated on both sides with very thin (essentially transparent) sheets of gold foil. The whole device was covered in a glass case, which allowed light to shine on the gold-selenium contact. Fritts found that the device successfully produced an electrical current with almost any kind and amount of light, ranging from full sunlight to the dim light of an incandescent bulb. In his report on his research, Fritts said that the current was "continuous, constant, and of considerable electromotive force" (Fritts 1885, 392). Unfortunately, it was not very efficient, converting no more than about 1 percent of the solar energy that fell on it into electrical energy. Still, Fritts held out high hope that his solar cell would challenge the coal-fired electric power plants developed by Thomas Edison only a few years before. In a later note about his work, he said that "we may ere long see the photo-electric plate competing with the dynamo-electric machine *[of Edison]* itself in a high percentage of electrical conversion (Siemens 1885, 515).

Fritts's work inspired a series of inventors over the next half century to find ways of modifying his basic solar cell design in order to improve its efficiency and, therefore, to make it economically feasible. Some of the inventors who followed Fritts's lead and applied for patents for solar cells include Edward Weston (1888), Melvin Severy (1894), Harry Reagan (1897), William Coblentz (1913), Anthony Lamb (1935), and Russell Ohl (1941). In no case, however, was any one of the patented devices adequate to meeting the challenge of a device that could efficiently convert solar energy into electricity ("Timeline of Solar Cells" 2015).

Like so many other discoveries in solar science, the work of Smith, Adams, Day, Fritts, and their colleagues went essentially nowhere, at least from a practical standpoint. The age of photovoltaics and solar cells was still nearly a half century off as the calendar turned to the year 1900. But the field of photoelectricity was by no means moribund at the theoretical level, as a number of researchers explored the fundamental question as to what was actually happening at the atomic level when light shines on a metal and releases electrons.

The problem was particularly interesting because the *way* in which light produces electrons from a metal does not occur according to classical laws of physics. According to those laws, increasing the intensity of light shined on a piece of metal should increase the flow of electrons released by the metal. But such is not the case. The electric current produced during the photoelectric effect is a function of the *frequency* (or wavelength) of the light used, not on its intensity.

The explanation for this unexpected result was provided in 1905 by Austrian American physicist Albert Einstein, an achievement for which Einstein was awarded the Nobel Prize in 1922. Einstein said that the photoelectric effect can be understood if one imagines that light consists of tiny individual particles, now known as *quanta* or *photons*, each with a certain amount of energy determined by its frequency (or wavelength, which is the inverse of the frequency). If particles of light have some minimum amount of energy (determined by their frequency), they can eject electrons from a metal surface; if they have less than that amount of energy, they cannot. At the time of this discovery, no one knew exactly how this new knowledge would have any effect on the solar power industry, although it is now known to be one of the theoretical cornerstones of solar science (Richmond 2015).

Solar Energy in the Twentieth Century

Technical Progress

By 1940, the earliest breakthroughs in the development of modern solar power systems were just beginning to appear. The first of those events occurred, as so often happens in science, as a chance discovery, by Bell Telephone Laboratories' researcher Russell Shoemaker Ohl. At the time, Ohl was studying the use of crystals for radio transmission and reception, trying to develop ultrapure crystals of silicon to achieve maximum efficiency for such applications. On one occasion, Ohl came across a silicon crystal that was anything but pure. In fact, it had a crack in it that seemed to disqualify it in his experiments. Before discarding the crystal, however, Ohl made an interesting discovery. Whenever light was shined on the crystal, a current flowed from one side of the crystal to the other. In demonstrating the effect to his coworkers at Bell, Ohl showed that simply shining a flashlight on the broken crystal produced an increase in potential across the crystal of half a volt, a significant amount under the circumstances ("Russell Ohl Accidentally Discovers the Silicon P-N Junction" 1999).

Before long, Ohl had devised an explanation for his strange result. He concluded that the silicon crystal contained some low level of impurities, but that the amount of impurities was somewhat different on either side of the crack. In effect, Ohl was holding in his hand a primitive version of a modern-day transistor, in which one side of the crystal contained an excess of negative charges (an N-type semiconductor) and the other side contained a deficiency of negative charges (a P-type semiconductor) ("This Month in Physics History: April 25, 1954: Bell Labs Demonstrates the First Practical Silicon Solar Cell 2009").

Ohl's discovery was the impetus needed to drive Bell's research into the direction that would result in the production of the modern solar cell. The next step in that process also occurred at Bell, through the joint efforts of three researchers there, Daryl Chapin, Calvin Fuller, and Gerald Pearson. In 1952, Chapin had been assigned the task of developing an alternative to traditional dry cell batteries, which did not function well in mild weather conditions. With a longtime interest in solar energy, Chapin decided to find out if solar cells might be the dry cell battery replacement for which Bell was looking. He began his research with selenium as the basis of his solar cells, at the time the only material that was in general use for such applications. He found out, however, that the best he could do with selenium was a solar cell that converted less than 1 percent of solar energy into electricity. That was certainly not going to satisfy Bell's needs.

About a year later, Chapin received some valuable input from friend and colleague Fuller, who, with Pearson, had been working on methods for doping the element silicon so as to convert it from a poor conductor into a good conductor. (The term *doping* refers to the intentional addition of small amounts of impurities to a material to change its electrical conductivity.) Fuller and Pearson had tried a number of possible dopants, including lithium, gallium, arsenic, and phosphorus, to improve the conductivity of silicon. In an early series of experiments, they found that a thin strip of silicon doped with arsenic and boron converted solar energy to electrical current with an efficiency of about 6 percent, at the lower limit of the dry cell battery alternative for which Bell had been looking (Perlin 2004). Bell Labs confirmed the success of the Chapin-Fuller-Pearson research in an announcement that the new device "compares favorably with the efficiency of steam and gasoline engines. Previous photoelectric devices have never rated higher than 1 percent" ("Solar Battery Turns Sunshine into Usable Electric Power" 1954, 71). In 1957, the three inventors received a patent for the invention of their solar cell (Solar Energy Converting Apparatus, US2780765A 1957).

In spite of its enthusiasm for the new discovery, Bell Labs and its commercial agents found it difficult to develop a business selling solar cells. The manufacture of the Chapin-Fuller-Pearson solar cell was still very expensive, such that electricity produced by such a cell cost about $300 per watt compared to a cost of about 50 cents a watt for electricity produced by conventional power plants. As a consequence, about the only applications found for solar cells were in devices such as toys, money-changing machines, and portable radios ("The History of Solar" 2015; Perlin 2015). Faced with this economic disparity, Chapin could hardly be blamed for wondering, "What to do with our new baby?" (Perlin 2005)

Fortunately, an answer to that question was in the offing. At just about the time that Chapin, Fuller, and Pearson were making their momentous discovery, another technological breakthrough was under way with the development of space vehicles. The date remembered as the dawn of the Space Age is October 4, 1957, when the Soviet Union launched the first space satellite, Sputnik 1. But space research had been going on long before that event, and the potential for solar power as an element in that program was not completely ignored. As early as 1951, U.S. researchers had begun to plan for their own space satellite, one that would be powered, however, by conventional chemical batteries. One person, German-born American physicist Hans Ziegler, however, had a different idea. (Ziegler was one of the German scientists who had been "encouraged" to move to the United States at the end of World War II.) Having seen a demonstration of the early Bell solar cell, Ziegler had become convinced that solar cells would provide a more reliable, if not necessarily less expensive, alternative to chemical batteries in space vehicles.

Ziegler's hopes, however, ran counter to those of the Vanguard contractor, the U.S. Navy, which regarded solar cells as "uncontrolled and not fully developed" (Perlin 2015). It was only because of delays in the Vanguard project and the launch of Sputnik 1 that the navy agreed to include solar cells as an "add-on" when the first space satellite, Vanguard 1, was launched on March 17, 1958. Ziegler's predictions eventually came true when the satellite's chemical batteries lasted only 19 days, while its solar cells continued to operate for another seven years, finally sending their last transmission in May 1964 ("Vanguard 1" 2014).

The story of solar cell development during the late 1950s and 1960s was one of slow, but certain, progress in design of the devices, on the one hand, and continued and expanded use for the powering of space vehicles on the other. As an example, Hoffman Electronics, one of the leading producers of solar cells during the period, gradually increased the efficiency of its product from about 8 percent in 1957 to 9 percent a year later and eventually to 14 percent by 1960. Also during the period, solar cells were chosen as the primary means of energy production on a host of space vehicles, including the Explorer VI and VII satellites launched in 1959, Telstar communications satellite (1962), Nimbus 1 (1964), the Orbiting Astronomical Laboratory (1966), Soyuz 1 manned spacecraft (1967), Soviet Salyut 1 (1971), and American Skylab 1 (1973) ("The History of Solar" 2015).

Solar Energy on the Upswing

During the 1970s, solar researchers continued to search for ways to increase the efficiency and reduce the cost of solar cells so as to extend their use for terrestrial purposes. But in 1973, a specific political event occurred that very quickly and dramatically changed the attitude of Americans and people in many other parts of the world about solar energy: the Arab oil embargo. While at war with the state of Israel, members of the

Arab Organization of Petroleum Exporting Countries (OPEC) decided to punish the United States and a handful of other nations for their support of Israel during that war. They chose to do so by significantly reducing their export of crude oil to those countries, thus vastly reducing the amount of gasoline and other petroleum products available and the going price for the products. In a single month, the price of oil increased from about $5 per barrel to more than $17 per barrel (Nersesian 2007, 165; "Oil Embargo, 1973–1974").

The oil embargo caused Americans to rethink their pattern of energy use, which for well over half a century had been predicated almost entirely on the belief that an abundant supply of coal, petroleum, and natural gas would "always" be available at inexpensive prices. One of the first U.S. reactions to the oil embargo was a decision by President Richard M. Nixon to initiate a new initiative to make the United States independent of foreign fuel supplies, a project that he called *Project Independence 1980*. The project depended heavily on the increased production of American fossil fuel supplies and improved conservation of energy resources, with only passing thought at first about the role that renewable energy could play in that effort (Nixon 1973).

One of the first specific actions taken to involve renewable resources in the new search for energy independence was the passage of the Solar Heating and Cooling Demonstration Act of 1974. This act took note of the fact that "the technologies for solar heating are close to the point of commercial application in the United States" and ordered the National Aeronautics and Space Administration (NASA) to begin a program of research to nudge this process of commercialization forward (Public Law 93–409 1974).

Nixon was unable to pursue his plan for energy independence since he resigned from the presidency less than a year after the speech in which he announced the plan. His successor, President Gerald Ford, however, continued to pursue much the same line on energy as did his predecessor. In one of his first major acts in office, Ford signed into law the Energy Reorganization Act of 1974, which dramatically changed the administrative structure of energy policy in the United States. One of the new agencies created by that act was the Energy Research and Development Administration (ERDA), which, among its other responsibilities, was responsible for all federal renewable energy projects. (ERDA itself remained in existence for only a short period, being replaced in 1977 by the Department of Energy Organization Act, which created the current U.S. Department of Energy.)

Solar energy was clearly in the minds of legislators as they began to work toward energy independence in 1974. Not only was solar mentioned in the Energy Reorganization Act itself, but two of the other four enabling acts also dealt with solar: the Solar Heating and Cooling Demonstration Act, mentioned earlier, and the Solar Energy Research Development and Demonstration Act of 1974. The latter act also included a key provision that established a new standalone research facility to be named the Solar Energy Research Institute (SERI), to be built in Golden, Colorado. (SERI was redesignated with a much broader charge as the National Renewable Energy Institute in 1991.)

Although Presidents Nixon and Ford and the congressional delegates with which they worked laid down the outlines of a reimagined national energy policy, concrete progress in this direction began to take place only with the election of President Jimmy Carter in 1976. Carter was the first president to focus his attention on the unique contributions that solar power could make in the nation's energy equation. The action for which he was perhaps best known was his decision to install solar panels on top of the White House, an act that occurred on June 20, 1979. However, Carter had much more ambitious goals for the development of solar power than the addition of 32 solar panels on the White House. The same evening of the panel installation, Carter spoke to the American people about his own views for the role of solar power in the United States in coming years. He said that "[w]e are today taking an historic step. We are making a commitment to as important a goal as we can set for our Nation—the provision of 20% of our energy needs from solar and renewable sources of energy by the year 2000" (Carter 1979).

In addition to his rhetoric about the importance of solar power to the United States, Carter listed a number of specific programs to achieve these objectives, including:

- an FY 1980 budget request for $1 billion for solar energy projects;
- the creation of a Solar Bank, designed to make loans for solar investments in residential and commercial buildings;
- an annual expenditure of $100 million from the Solar Bank to pay for such projects;
- tax credits for the construction of solar homes and commercial buildings;
- tax credits for agricultural and industrial projects that make use of solar energy;
- and a new series of solar demonstration projects managed by the Tennessee Valley Authority (TVA). (Carter 1979)

Some specific accomplishments resulting from the new emphasis on solar energy were apparent during the 1970s. For example, acting on its mandate under the Solar Heating and Cooling Demonstration Act of 1974, NASA designed and completed a project to bring low-cost solar power to 83 rural villages around the world. The solar power systems installed were used for a wide variety of purposes, such as water pumping, refrigeration, lighting, communication, food preparation, and the operation of cottage industries (Rosenblum et al. 1979, 4; Bifano, Ratajczak, and Martz 1979). NASA did not restrict this project to foreign countries but carried out similar projects in the United States, probably the best known of which took place on the Papago Indian Reservation at Schuchuli (Gunsight), Arizona. That system provided solar-powered electricity for 15 homes on the reservation, which had previously had no electrical connections. The electrical power was used for water pumping and home electricity until 1983, at which point the village was connected to the electrical grid and solar power was required only for water pumping (Bifano, Ratajczak, and Ice 1978).

Solar Energy: The Bottom Falls Out

By the beginning of the new decade, signs were beginning to appear that solar energy might finally be on the verge of becoming an economic reality in the United States. In 1982, the ARCO Solar company opened the world's largest commercial solar-powered power station in Hesperia, California. The plant had a rated capacity of 1 megawatt of power, but was being re-designed even as it opened with the goal of extending that number to 16 megawatts by 1984 (Arnett et al. 1983; "Solar Energy Shows Its Paces" 1983). But such signs soon turned out to be misleading. After a burst of enthusiasm for solar energy engendered by the oil embargo, interest in the new technology rapidly dissipated under the new presidential administration of Ronald Reagan. Reagan made his position on alternative energy clear in a presidential debate with then-president Carter on October 28, 1980. He expressed the view that research on alternative energy programs in the preceding decade had been a waste of money. "It hasn't produced," he said, "a quart of oil or a lump of coal, or anything else in the line of energy" ("October 28 1980 Debate Transcript" 2014).

Once elected, Reagan was true to this view of alternative energy. In a relatively short period of time, he cut the Department of Energy's budget for alternative energy and conservation by half, reduced the budget of National Renewable Energy Laboratory (NREL) by nearly 90 percent, eliminated the wind investment tax credit (more about this item in the next chapter), reduced funding for research on solar photovoltaic cells by two-thirds, and removed the solar panels on the White House

(Parry 2014). The last of these actions was almost certainly symbolic only, with essentially no effect at all on the White House energy budget. Carter had taken the action almost entirely to symbolize his commitment to the development of alternative energy sources. Seven years later, essentially without comment, Reagan ordered the panels to be removed—again, an act apparently designed for symbolic, rather than practical, purposes (Biello 2010). (Solar panels were once more installed on the White House during the administration of President Barack Obama in 2010. See Eilperin 2014; Wihbey 2008.)

The concrete result of Reagan's view on renewable energy (and that of his successor, President George H. W. Bush) can be seen in budget allocations for this line of research over the next 12 years. The average annual appropriation for research on renewable energy during the Carter administration (FY 1978–FY 1981) was $1.290 billion. During the eight-year Reagan administration (FY 1982–FY 1990), the average annual appropriation fell to $253 million, and under Bush (FY 1991–FY 1995), it fell slightly again to $209 million annually ("R&D Priorities within the Department of Energy" 2015, Table IV.2.2).

Solar energy suffered sharp cuts especially under the Reagan administration. In his proposed budget for FY 1982, Reagan recommended reducing outlays for solar research from the projected expenditures under the Carter administration by as much as about 60 percent, as shown in Table 1.1. Particularly hard hit was SERI, which President Carter had seen as the flagship research institute for solar energy in the coming years. Shortly after Reagan took office, the institute lost a significant portion of its funding, along with important members of staff and assurance of a long-term (or even permanent) part in the nation's development of solar energy. Reagan's budget explained some of the cutbacks for SERI because "[t]he mission of this organization needs to be better defined and an appropriate staffing level agreed upon before any decision is made to proceed" ("Fiscal Year 1982. Budget Revisions. Additional

Table 1.1   Budget Requests for Solar Research, 1981–1986 (millions of dollars).

Year	Carter Budget	Reagan Budget	Change (%)
1981	577	478	–17
1982	583	193	–67
1983	664	207	–69
1984	623	220	–65
1985	595	233	–61
1986	553	244	–56

Source: "Fiscal Year 1982. Budget Revisions. Additional Details on Budget Savings. April 1981." Federal Reserve Archive. https://fraser.stlouisfed.org/docs/publications/usbrev/bus_rev_1982_v2.pdf. Accessed on February 25, 2015.

Details on Budget Savings. April 1981"). The impact on the institute was described some years later by Denis Hayes, who had been chosen by Carter to be the "solar czar" for research in the United States. In an interview some years after the transition from Carter to Reagan, Hayes said that "[i]n June or July of 1981, on the bleakest day of my professional life, they descended on the Solar Energy Research Institute, fired about half of our staff and all of our contractors, including two people who went on to win Nobel prizes in other fields, and reduced our $130 million budget by $100 million" (Koff 2005).

Implementing Solar Power

For all the technical developments in the field of solar energy and all the intense political and economic debates about the resource, one basic question that could be asked is, what difference did all of this make in terms of the use of solar power in the United States and the rest of the world? That is, was more solar power being produced in the second half of the twentieth century, and was solar energy becoming a more important part of the nation's energy equation?

The answer to those questions can be found in three basic measures of solar energy: (1) how efficient was it, especially in comparison to conventional fossil fuels? (2) how much did it cost? and (3) how many consumers were using solar energy instead of (or in addition to) coal, oil, and natural gas?

As noted earlier, the efficiency of solar cells increased gradually over the last third of the twentieth century. Following the trends in solar cells, efficiency has become somewhat complicated because so many different kinds of cells are now available, including the following general types:

- Silicon cells, one of the first types of solar cells to be developed in which the working medium is silicon;
- CIGS cells, which contain the four elements copper, indium, gallium, and selenide;
- Cd/Te cells, which contain the elements cadmium and tellurium as the primary working components;
- III-V cells, which contain elements in groups III, IV, and V in the periodic table; and
- OPV cells, which make use of organic compounds (the "O" in OPV) to make photovoltaic (the "PV" in OPV) cells.

A very useful review of the changes in solar cell efficiencies can be found at the report of a lecture on progress in solar cells given by Gregory M. Wilson of the National Renewable Energy Laboratory (NREL) in 2013. The report includes a timeline of the most efficient solar cells of each type from about 1975 to 2013. As this chart shows, the efficiency of the oldest type of solar cell, the amorphous silicon solar cell (a-Si), has increased from just above 0 percent in 1975 to more than 12 percent in 2013 (Wilson 2013). A more efficient type of silicon solar cell, the crystalline silicon solar cell (c-Si), was first developed in the early 1940s, but its efficiency at that point was only about 1 percent. Very rapid progress was made in its design in the late 1960s, and it soon reached an efficiency level of about 14 percent in the late 1970s (Green 1993).

III-V solar cells were first developed in the late 1970s and had an initial efficiency of about 16 percent. Innovative designs very quickly improved that value until, in the 2010s, such cells had become the most efficient of all solar cells currently available, with a value of more than 40 percent in 2013 (Cotal et al. 2009; Wilson 2013). Organic photovoltaic cells (OPV) were first developed in the mid-1970s, when, as with other new solar cells, they had a very low efficiency of about 1 percent. Progress in OPV design took place very slowly, and by the early 2000s, their efficiency was still much less than 10 percent. More recent progress, however, has resulted in a significant increase in efficiency, at a rate faster than that for any other type of solar cell, reaching about 10 percent in 2013 (Abdulrazzaq et al. 2013; Wilson 2013; for a more technical and detailed discussion of solar cell efficiencies, along with best values currently available with various types of cells, see "The Efficiency of Solar Panels" 2012).

These data make it clear that innovations over the past half century have significantly increased the efficiency of solar energy systems. However, those systems are still generally much less efficient than traditional, or even other renewable, energy sources. For example, according to studies conducted by Eurelectric, an association of European electric companies, the most efficient types of energy systems are those that make use of running water, hydroelectric and tidal system in particular. These systems typically have efficiencies in excess of 90 percent. That is, at least 90 percent of the energy contained in the running water is converted to another form, almost always electrical energy. Coal- and oil-fired power plants are about half that efficient, ranging in efficiency from about 40 to 45 percent. That is, half of the energy stored in the coal or oil is converted into electricity; the rest is lost to the plant surroundings in the form of heat. Nuclear fission plant efficiency falls close to the range for fossil fuel plants. Renewable energy sources are generally still not very efficient: wind turbines and biomass, about 30 percent; solar thermal, about 15 percent; photovoltaic, about 12 percent; and geothermal, about 12 percent ("Energy Efficiency" 2015).

The conclusion of this review is that the efficiency of solar cells has greatly improved, especially in the last two decades, but that they are not yet able to challenge the efficiency of other types of energy resources, especially coal, oil, natural gas, and nuclear.

A second fundamental question has to do with the price of solar cells and other solar equipment. For most of the history of solar energy in the modern era, solar power has not been able to compete with fossil fuels, nuclear power generation, and even other forms of renewable energy simply because the equipment needed for the capture of solar energy was too expensive to compete with these other forms of energy. For decades, proponents of solar energy have argued that if the price of solar cells and other solar equipment could be reduced sufficiently, solar energy could then compete with traditional energy sources.

Data and statistics on this question are somewhat difficult to summarize because the cost of different types of solar equipment varies considerably. For example, the cost of silicon solar cells is significantly different from organic photovoltaic devices or other types of solar cells. One indication of the trends in cost of solar equipment, however, is the price of the typical solar module over a period of years.

A brief aside about terminology first:

A **solar cell** is the smallest discrete unit used for the conversion of solar energy to electricity. More about the construction of a solar cell later in this chapter. A **solar module** consists of a group of solar cells, typically about 40 cells, connected to each other to increase the flow of electrical current from the system. A **solar panel** consists of a collection of solar modules and can have anywhere from two to two dozen or more modules per panel. Finally, a **solar array** consists of a collection of solar panels that can be of virtually any size at all, depending on the demands of a specific facility ("Renewable Energy Fact Sheet: Solar Cells" 2013; for an interesting demonstration of the way solar arrays are made, see "How Solar Panels Are Made" at http://www.yinglisolar.com/us/solar-basics/).

In addition, surveys may report on the cost of solar energy, including all installed costs or just for operation of the system, or the cost per unit of electrical power, such as dollars or cents per watt or per kilowatt. One report on the trend of photovoltaic solar cells, for example, indicated that the price of a single solar module dropped from nearly $24 per unit in 1980 to about $4 per unit in 2000 to less than a dollar per module in 2013 (Baldwin and Dowd 2013, 25). Another study found that the price of the average solar module per watt dropped from about $23 per watt in 1980 to just over $2 per watt in 2010 ("2010 Solar Technologies Market Report" 2011, 60; for other trend data, see also; Ardani 2012, 5; "The Increasing Efficiency of Renewable Energy" 2015; Mints 2013; "Solar Electricity Costs" 2015; "Solar Stats" 2012).

The bottom line in this discussion probably comes down to one question: how many solar cells, modules, panels, and arrays are being built, sold, and used? Interestingly enough, not a great deal of information is available on this question prior to the Carter presidency, when most surveys lumped all forms of renewable together. That is, so little energy was being produced by solar, wind, geothermal, tidal, and other renewable sources that federal agencies and other surveying agencies simply asked about and reported on "renewable energy" in general. The first year in which the federal government was able to report specific numbers for solar energy production and consumption was 1985, when the Energy Information Administration (EIA) reported that less than 0.5 trillion Btu of solar energy was consumed in the United States. By contrast just over 6,000 trillion Btu of all forms of renewable energy was consumed, almost all in the form of hydroelectric power and biomass conversion ("Monthly Energy Review" 2015, Table 10.1, 147).

After 1985, production and consumption of solar energy began to grow, but at a very slow pace and in an amount that still qualified as only a few percent (or less) of total energy production and consumption in the United States. Table 1.2 shows how these values changed for the production of electrical

Table 1.2 Electrical Generation by Solar Power, 1990–2014 (million kilowatt-hours)

Year	Solar	Percentage of all Electrical Power	Percentage of all Renewable Energy for Electrical Power
1990	367	0.012	0.12
1995	497	0.012	0.13
2000	493	0.013	0.13
2001	543	0.014	0.19
2002	555	0.014	0.16
2003	534	0.014	0.15
2004	575	0.014	0.16
2005	550	0.014	0.15
2006	508	0.012	0.13
2007	612	0.015	0.17
2008	864	0.021	0.23
2009	891	0.022	0.21
2010	1,212	0.029	0.28
2011	1,818	0.044	0.35
2012	4,327	0.105	0.87
2013	9,036	0.228	1.78
2014	18,321	0.448	3.39

Source: "Monthly Energy Review." June 2015. 2015. Energy Information Administration, Table 7.2a, 105. http://www.eia.gov/totalenergy/data/monthly/pdf/mer. pdf. Accessed on February 26, 2015. Percentage values for columns 3 and 4 calculated from table values by author.

power from 1990 to 2014, the last year for which data are available. (The use of solar energy for the production of other forms of energy is still too low to appear in most energy consumption tables in the United States ["Monthly Energy Review 2015," Table 7.2a, 105 and Table 2.4, 33].)

The numbers shown here might be discouraging for a proponent of solar energy, except that they do not really tell the complete story about the role of solar power in the United States and world energy equations. Far more important than the raw numbers themselves are the trends in the construction of solar energy facilities and the rapid increase in production and consumption of solar power. Data and graphs from the Department of Energy's Office of Energy Efficiency & Renewable Energy *2013 Renewable Energy Data Book* provide a much more optimistic view of the future of solar energy. According to that resource, solar photovoltaic power generating *capacity* has increased by more than 50 percent annually in 11 of the 13 years between 2000 and 2013. The only other renewable resource that can match those numbers is wind, which has increased its capacity by more than 50 percent in two of those years (2013 Renewable Energy Data Book 2014, 22). Similarly, the increase in actual electrical energy *production* by solar power has risen from less than 10 percent annually in the period between 2000 and 2004 to 54 percent in 2010, 66 percent in 2011, 71 percent in 2012, and 66 percent in 2013, the last year for which data are available. Again, only wind power comes even close to these numbers, with its best year in 2008 at 60.7 percent annual increase and less than 30 percent throughout most of the rest of the period (2013 Renewable Energy Data Book 2014, 28).

These patterns are reflected in worldwide data, which show trends similar to those in the United States. For example, the Energy Data Book statistics show that the annual percentage increase in solar power capacity is far greater than that for any other energy source. Solar photovoltaic capacity, for example, increased worldwide by 90 percent in 2010, 78 percent in 2011, 41 percent in 2012, and 39 percent in 2013. Comparable rates for wind were 25 percent, 20 percent, 19 percent, and 12 percent, and for other renewable sources, generally less than 10 percent. Even more interesting has been the increase in concentrated solar power (CSP) facilities of 83 percent, 43 percent, 57 percent, and 36 percent worldwide in 2010, 2011, 2012, and 2013, respectively (2013 Renewable Energy Data Book 2014, 43). Numbers such as these can only bring good feelings to those who see a bright future for solar energy in the United States and the world.

An Introduction to Solar Technology

Solar energy systems can be subdivided into one of three major types of systems: space heating and cooling (SHC), photovoltaic cells (PVC), and concentrating solar power (CSP).

Space Heating and Cooling (SHC)

The oldest of these systems, space heating and cooling, has already been discussed in some detail earlier in this chapter. SHC systems can be further subdivided into two general categories: passive systems, such as those described earlier, and active systems, in which some type or types of mechanical devices are used to collect solar energy and convert it to heat energy and then move that energy to places where it can be used. Recall that passive systems rely essentially on the way a building is constructed, orienting walls and roofs in such a way as to collect solar energy during the winter, when it can be used to heat a building, and to deflect sunlight in the summer, as a way of keeping a building cool.

An image of a typical active SHC can be found at the "What Is Solar Energy" website (http://www.theecoambassador.com/solarenergy.html) (among many other sources). In a system of this type, a solar array is used to collect solar radiation, which is then used to heat a working fluid. The working fluid can be either a gas (such as air) or a liquid (such as water). Pumps move the heated fluid through pipes first into a storage container, where it is kept warm during evening hours when there is no sunlight. Other pumps then deliver the heated fluid from the storage tank to locations within a building where it can be used, such as the kitchen or bathroom. Some SHC systems also have a heat exchange unit where the heated fluid can be used to heat a second working fluid for other applications.

SHC is used in a variety of industrial, commercial, and residential settings such as private homes and multifamily residences, hotels and motels, car washes, nursing homes, hospitals, laundries, food service and processing facilities (such as restaurants and breweries), residential and public swimming pools and hot tubs, and similar sites where a plentiful supply of hot water is needed over extended periods of time (Dutzik, Kerth, and Sargent 2011, 10–13).

Photovoltaic Cells (PVC)

The photovoltaic cell (PVC) has also been described in some detail earlier in this chapter. As a review, see a simple diagram of a solar cell such as the one on the "How Do Solar Panels Work?" website at http://www.redarc.com.au/solar/about/solarpanels/. The "working part" of a solar cell is internal bilayer consisting of an N-type semiconductor and a P-type semiconductor. When sunlight strikes the system, negative charges (electrons) flow through the bilayer in one direction and positive charges ("holes") flow in the other direction. The current thus produced then leaves the solar cell and is transmitted to a device in which it is used. Today, a great variety of semiconductor systems are available, differing primarily in the types of materials used in their construction. (For an excellent review of the types of solar cells available now, see Hermann 2011.)

The actual construction of a solar cell is somewhat more complicated than many diagrams show, however, because the cell must be protected from damage and connected to other cells. A more detailed image of the parts of a complete solar cell (which is sometimes, but rarely, also called a solar module) can be found at "Photovoltaics: Solar Cells 101" at http://alpenglowsolar.com/solar-cells-101.php and "Solar Energy 101: Introduction to Solar Energy" at http://www.dowcorning.com/content/solar/solarworld/solar101.aspx.

Concentrating Solar Power (CSP)

A third basic type of solar technology is concentrating solar power (CSP), which uses some sort of device to focus sunlight into a narrow region, where it is used to raise the temperature of a working fluid (or heat transfer medium). A great number of materials have been tested as working fluids, including water (steam), organic compounds and mixtures, molten salts and metals, sand, ionic liquids, and carbon nanotubes. The heated fluid is then transferred to turbines that operate generators that produce electricity. The system is basically similar to that of a coal-, oil-, or gas-fired power plant, except that the source of energy that changes water into steam in the system is sunlight rather than a fossil fuel.

Today, four basic types of CSP systems are in operation: parabolic trough, tower power, solar dish, and linear Fresnel plants. The four types differ from each other primarily in the way they focus (or concentrate) solar radiation. The parabolic trough system makes use of long curved mirrors with pipes running through the center of the parabolic troughs. The mirrors bring the Sun's ray to a focus at the center of the parabolic curves, where the pipes are located. The working fluid inside the pipes is heated and then pumped to the turbine and generator or to a holding tank where it is kept warm for future use. (For a diagram of the system, see Hamilton 2015, Diagram 2.)

In a second type of CSP system, a tall tower is surrounded by a number of flat mirrors (called *heliostats*) pointed at a receiver at the top of the tower. The working fluid passes into the receiver, where it is heated, and then pumped out of the tower into a turbine and generator and/or a collecting tank (Hamilton 2015, Diagram 3). The solar dish system consists of a group of mirrors that look like home-satellite dishes with collecting vessels at the focal points of the mirrors. (The focal point of a curved mirror is the point at which all the rays reflected from the mirror come together.) The working fluid inside the collecting vessels is heated by the concentrated solar radiation, passes downward through a series of pipes and onto a turbine, generator, and/or storage tank (Hamilton 2015, Diagram 4).

The most recent type of CSP system is the linear Fresnel system, which is very similar to the parabolic trough system, except that the parabolic mirrors in the latter are replaced by flat mirrors, which are simpler and less expensive to make. Sunlight shining on the mirrors is reflected upward onto a set of pipes containing the working fluid, which is then used to drive a turbine and generator, as in other systems (see Hamilton 2015, Diagram 5; for an excellent general overview of CSP, see "Concentrating Solar Power: Technology Brief" 2013).

Thin-Film and Hybrid Systems

Many variations on these basic types of solar technology exist. Two that are worthy of special mention are thin-film photovoltaics and hybrid solar systems. Thin-film photovoltaics are very thin solar cells (anywhere from a few nanometers to a few micrometers in thickness) laid down on flexible strips of plastic. The working material in these devices can be cadmium telluride (CdTe), copper indium gallium selenide (CIGS), or some form of amorphous silicon. Thin-films are generally less expensive than more traditional types of solar cells, but they have only recently become as efficient as those devices. They currently constitute about 10 percent of the international market in solar cells and have an efficiency that ranges from about 20 percent for CdTe and CIGS cells to 25 percent for some types of silicon cells ("Photovoltaics Report" 2014, 4, 6).

The term *hybrid solar system* can have a variety of meanings. In all cases, however, they refer to the fact that more than one type of (usually) renewable energy is used in the system. For example, one type of hybrid system combines electricity produced by both wind and solar. One such system might be a home with a small wind turbine and a solar array attached to its roof. During the summer, when solar energy is more readily available than wind, the solar array generates most of the electricity the house needs. In the winter, when the amount of solar energy is reduced, but winds are stronger, wind energy is responsible for most of the home energy needs ("Hybrid Energy" 2015).

Another type of hybrid solar system is one in which solar radiation is used to generate both electrical current (through a photovoltaic system) and hot water (through a solar thermal system). In this system solar radiation is captured by a solar array on the roof of a house, where an electrical current is produced. Some of the current goes directly to end uses in the house, such as lighting the building and operating appliances. Some of the current also goes to solar thermal systems, where the electricity is used to heat water for use in the house ("Solar Power Products" 2015).

Space-Based Solar Power (SBSP)

One of the most promising and exciting forms of solar technology that may be possible is space-based solar power (SBSP), also known as space solar power (SSP), solar-power satellite (SPS), and satellite power system (SPS). An SBSP system consists of devices that collect and concentrate solar radiation in outer space and then transmit that radiation to Earth, where it is converted into electricity. A number of models have been proposed since the concept was first proposed in the 1970s. Some possible designs are shown on the web page for the blog Billion Year Plan at http://billionyearplan.blogspot .com/2011_09_01_archive.html. Even the most optimistic observers say that a working SBSP system may be at least a decade away, but they also say that such systems are likely to be inevitable given the many advantages of space-based solar energy. Such advantages include the fact that an SBSP system takes up no space on Earth's surface, emits no waste products to the atmosphere, operates 24 hours a day and seven days a week year-around without interruption, allows any and every nation that can afford the system to become energy independent, and makes it possible to provide electrical energy to any point on Earth, no matter how remote it may be. The main disadvantage, of course, is the cost of constructing such a system and, perhaps, maintaining it in case of breakdowns ("Space Solar Power" 2014).

As of mid-2015, a number of nations are making fairly specific plans for the construction of satellite-based solar power systems. The Chinese, for example, have announced plans for building an SBSP system that will be operational by 2040, and the Japanese have designed a similar system that will supply electricity to 300,000 homes in that energy-poor nation (Rajagopalan and Prakash 2011; Edwards 2009; an introduction to the U.S. SBSP can be found at Wood 2014).

Solar Energy Uses

In the broadest possible sense, solar energy can be said to have two primary uses: to heat or cool buildings and other spaces and to generate electricity. That general description includes a number of more specific applications, such as heating water systems in homes and commercial buildings; heating swimming pools and spas; providing heat for homes and other buildings; operating fans that can cool and ventilate structures; operating water pumps and other types of pumps in a home, commercial building, or industrial structure; charging batteries; providing the electrical energy needed for operating a vast array of machines and other devices found in residential buildings, commercial buildings, and industrial factories; and running indoor and outdoor lighting systems (Thiele 2015).

But solar energy has a number of other applications also, many not well known, particularly popular or in widespread use, or much beyond the trial stage. For example, a number of inventors have long been studying the use of solar energy to operate a variety of forms of transportation, including cars, buses, and aircraft. The first solar-powered car is thought to have been the 15-inch-long Sunmobile, designed by General Motors engineer William Cobb in 1955 ("William Cobb Demonstrates First Solar-Powered Car" 2014). In the six decades since that first step forward, solar-powered-car technology has progressed very slowly and, for the most part, has consisted of "proof of concept" experiments, designed to show that such cars are even possible. Most of the cars produced in this research have been built to participate in national or worldwide competitions of solar cars, such as the American Solar Challenge and the World Solar Challenge. According to the Guinness World Book of

Records, the current world speed record for vehicles powered entirely by solar energy is 91.332 miles per hour, registered in 2014 by Kenjiro Shinozuka, driving the Sky Ace TIGA, built by students at Japan's Ashiya University ("Fastest Solar-Powered Vehicle" 2015).

So far, relatively few automotive companies have invested in the development of a solar-powered car for commercial use. One of the very few exceptions to that trend is the Ford Motor Company, which, in 2014, released to the public its first solar-powered vehicle, the C-Max Solar Energi Car. The car is a hybrid that operates on a gasoline engine and battery powered by solar energy collected by a solar array on its roof. It seats five passengers and has a maximum speed of 115 miles per hour ("The Advanced Technology, Fun-to-Drive Hybrid" 2015).

As with land vehicles, the first solar-powered aircraft were built in the 1970s. AstroFlight Sunrise I was a small (27-pound) unmanned aircraft that flew for the first time at Fort Irwin Military Reservation in November 1974. It was built under an experimental contract issued by the U.S. Department of Defense. Five years later, the first manned solar-powered spacecraft, Mauro Solar Riser, flew out of Flabob Airport in Riverside, California, in April 1979 (Noth 1980, 4–6). Since that time, a large number of experimental solar-powered aircraft have been designed, built, and tested. One of the most promising aircraft yet built is Solar Impulse, built by a privately financed company led by Swiss psychiatrist and aeronaut Bertrand Piccard and Swiss businessman André Borschberg. Initial feasibility studies were conducted in 2003 at the École Polytechnique Fédérale de Lausanne, and the first manned flight was completed in 2009. Solar Impulse's greatest accomplishment thus far has been an around-the-world flight conducted in 2015 out of Abu Dhabi (Solar Impulse Beta 2015). The final leg of that flight, from Hawaii to Arizona, was postponed in mid-2015 because of damage to the plane's batteries on the flight from Japan to Hawaii. The flight is now expected to be completed sometime in early 2016.

Conclusion

In some respects, solar technology seems as if it has hardly advanced much at all in 2,000 years. Architects and builders today are still designing homes and other buildings to take maximum use of solar radiation to keep the building warm in winter and cool in summer and to heat water for cooking, cleaning, industrial operations, and other purposes. Solar engineers are still looking for ways to capture solar radiation and convert it into heat (as in concentrating solar power systems) or electricity (as in photovoltaic systems).

But, of course, such comparisons have only limited value. Modern solar technology, even if it uses the same basic principles as those employed by the early Egyptians and ancient Greeks and Romans, is far more advanced and sophisticated than those early predecessors. The most significant advances in solar technology have, however, occurred only in the past 50 years ago. So proponents of solar technology can be encouraged by the fact that the Solar Age truly seems barely to have gotten under way. It is only in the second decade of the twenty-first century that the true potential of solar energy is becoming clear. And with that change has also come a number of issues as to how the world can take the best advantage of this limitless resource and how it will be able to deal with some of the many social, political, economic, environmental, and other problems that have arisen and will continue to arise as solar power becomes even more widely popular in the United States and around the world.

References

Abbott, Charles Greeley. 1938. *The Sun and the Welfare of Man*, vol. 2. New York: Smithsonian Scientific Series. http://digicoll.library.wisc.edu/cgi-bin/HistSciTech/ HistSciTech-idx?type=turn&entity=HistSciTech.SunAbbot .p0286&id=HistSciTech.SunAbbot&isize=text. Accessed on February 16, 2015.

Abdulrazzaq, Omar A., et al. 2013. "Organic Solar Cells: A Review of Materials, Limitations, and Possibilities for Improvement." *Particulate Science and Technology*. 31(5): 427–442.

Adams, W. G., and R. E. Day. 1876. "The Action of Light on Selenium." Proceedings of the Royal Society of London. 25: 113–117. http://rspl.royalsocietypublishing .org/content/25/171-178/113.full.pdf. Accessed on February 16, 2015.

"The Advanced Technology, Fun-to-Drive Hybrid." 2015. Ford. http://www.ford.com/cars/cmax/features/?search id=171888534|10523849814|86742700854&ef_id=V L77NwAABeX1pYo8:20150301175439:s. Accessed on March 1, 2015.

Ardani, Kristen. 2012. "Soft Costs in the U.S. Solar Markets—Survey Results." National Renewable Energy Laboratory. http://votesolar.org/wp-content/ uploads/2012/11/NREL_SoftCost_Webinar.pdf. Accessed on February 26, 2015.

Aristophanes. 419 BCE. "The Clouds." The Internet Classics Archive. http://classics.mit.edu/Aristophanes/clouds.html. Accessed on February 10, 2015.

Arnett, J. C., et al. 1983. "Design, Installation and Performance of the ARCO Solar One-Megawatt Power Plant." In Willeke Palz and F. Fittipaldi, eds. *Photovoltaic Solar Energy Conference; Proceedings of the Fifth International Conference, Athens, Greece, October 17–21, 1983.* Dordrecht: D. Reidel Publishing Company, 314–320.

"Arnold Schwarzenegger: Green Quest Goes On." 2012. BBC News. http://www.bbc.com/news/ world-us-canada-17863391. Accessed on February 5, 2015.

Baldwin, Sam, and Jeff Dowd. 2013. "Energy Efficiency & Renewable Energy." http://ip-science.thomsonreuters .com/m/pdfs/fed-res/baldwin_talk_tr_2013-03-22.pdf. Accessed on February 26, 2015.

Biello, David. 2010. "Where Did the Carter White House's Solar Panels Go?" Scientific American. http://www .scientificamerican.com/article/carter white house solar panel array/. Accessed on February 25, 2015.

Bifano, William J., Anthony F. Ratajczak, and James E. Martz. 1979. "A Photovoltaic Power System in the Remote African Village of Tangaye, Upper Volta." National Aeronautics and Space Administration. Lewis Research Center. http://ntrs.nasa.gov/archive/nasa/ casi.ntrs.nasa.gov/19800004300.pdf. Accessed on February 19, 2015.

Bifano, William J., Anthony F. Ratajczak, and William J. Ice. 1978. "Design and Fabrication of Photovoltaic Power System for the Papago Indian Village of Schuchuli (Gunsight), Arizona." National Aeronautics and Space Administration. Lewis Research Center. http://ntrs.nasa .gov/archive/nasa/casi.ntrs.nasa.gov/19780018612.pdf. Accessed on February 19, 2015.

"Billion Year Plan." 2015. http://billionyearplan .blogspot.com/2011_09_01_archive.html. Accessed on July 22, 2015.

Botkin, Daniel. 2010. "Powering the Future. A Scientist's Guide to Energy Independence." http://ptgmedia .pearsoncmg.com/images/9780137049769/ samplepages/9780137049769.pdf. Accessed on February 8, 2015.

Butti, Ken, and John Perlin. 1980. *A Golden Thread: 2500 Years of Solar Architecture and Technology.* Palo Alto, CA: Cheshire Books. http://www.arvindguptatoys .com/arvindgupta/golden-thread.pdf. Accessed on February 7, 2015.

Cahill, Nicholas. 2008. *Household and City Organization at Olynthus.* New Haven, CT: Yale University Press. http://www.stoa.org/olynthus/. Accessed on February 8, 2015.

Cardwell, Diane. 2015. "Solar Power Battle Puts Hawaii at Forefront of Worldwide Changes." New York Times. April 19, 2015. http://www.nytimes.com/2015/04/19/business/energy-environment/solar-power-battle-puts-hawaii-at-forefront-of-worldwide-changes.html?_r=0. (Subscription required) Accessed on April 20, 2015.

Carter, Jimmy. 1979. "Solar Energy Message to Congress. The American Presidency Project." http://www.presidency.ucsb.edu/ws/?pid=32503. Accessed on February 19, 2015.

Caus, Salomon de. 1615. *Les raisons des forces mouvantes.* Francfort: Jan Norten. http://cnum.cnam.fr/ILL/FDA1.html. Accessed on February 10, 2015.

Church, William Conant. 1890. *The Life of John Ericsson.* London: Sampson Low, Marston, Searle & Rivington.

Cleveland, Cutler J., and Christopher Morris, eds. 2015. *Dictionary of Energy*, 2nd ed. Amsterdam: Elsevier.

Clinton, Hillary Rodham. 2005. "Remarks of Senator Hillary Rodham Clinton to the Cleantech Venture Forum VIII." http://www.esf.edu/energycenter/bioproeng/2005/10.25.NEB%20Senator%20Clinton%20Energy%20Speech.htm. Accessed on February 5, 2015.

Collins, Paul. 2002. "The Beautiful Possibility." *Cabinet.* http://www.cabinetmagazine.org/issues/6/beautifulpossibility.php. Accessed on February 13, 2015.

"Concentrating Solar Power: Technology Brief." 2013. International Renewable Energy Agency and the Energy Technology Systems Analysis Programme. http://www.irena.org/DocumentDownloads/Publications/IRENA-ETSAP%20Tech%20Brief%20E10%20Concentrating%20Solar%20Power.pdf. Accessed on March 2, 2015.

Cotal, Hector, et al. 2009. "III–V Multijunction Solar Cells for Concentrating Photovoltaics." *Energy & Environmental Science*. 2(2): 174–192. http://www.spectrolab.com/pv/

support/Cotal_III_V_multijunction_photovoltaics.pdf. Accessed on February 25, 2015.

Czochralski, J. 1918. "Ein neues Verfahren zur Messung der Kristallisationsgeschwindigkeit der Metalle" ("A New Method for the Measurement of the Crystallization Rate of Metals"). *Zeitschrift für Physikalische Chemie*. 92: 219–221. (In German)

Di Pasquale, Giovanni. 2004. "Scientific and Technological Use of Glass in Graeco-Roman Antiquity." In Marco Beretta, ed. *When Glass Matters: Studies in the History of Science and Art from Graeco-Roman Antiquity to Early Modern Era*. Firenze, Italy: Leo S. Olschki.

Dutzik, Tony, Rob Kerth, and Rob Sargent. 2011. "Smart, Clean and Ready to Go: How Solar Hot Water Can Reduce Pollution and Dependence on Fossil Fuels." PennEnvironment Research and Policy Center. http:// www.pennenvironment.org/sites/environment/files/ reports/Smart-Clean-and-Ready-to-Go_.pdf. Accessed on February 27, 2015.

Edwards, Lin. 2009. "$21 Billion Orbiting Solar Array Will Beam Electricity to Earth." Phys.Org. http://phys.org/ news172224356.html. Accessed on March 1, 2015.

"Effect of Light on Selenium during the Passage of an Electric Current." 1873. *Nature*. 7(173): 303. http://digicoll.library.wisc.edu/cgi-bin/HistSciTech/ HistSciTech-idx?type=div&did=HISTSCITECH. NATURE18730220.SMITH01&isize=M. Accessed on February 17, 2015.

"The Efficiency of Solar Panels." 2012. Green Energy. http:// greenenergy-power.com/the-efficiency-of-solar-panels/. Accessed on February 28, 2015.

Eilperin, Juliet. 2014. "Solar Panels Here to Stay Atop White House Roof." The Washington Post. http://www .washingtonpost.com/blogs/post-politics/wp/2014/05/09/

solar-panels-here-to-stay-atop-white-house-roof/?wprss
=rss_homepage&clsrd. Accessed on February 25, 2015.

"Energy Efficiency." 2015. Electropaedia. http://www
.mpoweruk.com/energy_efficiency.htm. Accessed on
February 26, 2015.

"Engine Power from Solar Heat." *Scientific American*.
(Reprint of a 1909 article). http://www.scientificamerican
.com/article/engine-power-from-solar-heat/. Accessed on
February 16, 2015.

"Fastest Solar-Powered Vehicle." 2015. Officially
Amazing. http://www.guinnessworldrecords.com/
world-records/fastest-solar-powered-vehicle. Accessed on
March 1, 2015.

"First Photovoltaic Devices." 2015. PV Education.org.
http://pveducation.org/pvcdrom/manufacturing/
first-photovoltaic-devices#footnote1_zbrx8nx. Accessed on
February 16, 2015.

"Fiscal Year 1982. Budget Revisions. Additional Details on
Budget Savings. April 1981." Federal Reserve Archive.
https://fraser.stlouisfed.org/docs/publications/usbrev/bus_
rev_1982_v2.pdf. Accessed on February 25, 2015.

Fritts, C. E. 1885. "On the Fritts Selenium Cell and
Batteries." *Van Nostrand's Engineering Magazine*. 32:
388–395. https://books.google.com/books?id=7gUTAAA
AYAAJ&pg=PA514&lpg=PA514&dq=%22on+the+electr
omotive+action+of+illuminated+selenium%22&source=bl
&ots=bYJPZp4x_F&sig=NlFdmyh1WMkPoJ-bFmRh10w
bB8Y&hl=en&sa=X&ei=k47iVLT-J83boATQ_oDICg&v
ed=0CC4Q6AEwBQ#v=snippet&q=continuous%2C%20
constant&f=false. Accessed on February 16, 2015.

Green, Martin A. 1993. "Silicon Solar Cells:
Evolution, High-Efficiency Design and Efficiency
Enhancements." *Semiconductor Science and Technology*.
8: 1–12. http://copilot.caltech.edu/classes/aph9/

IoP_Martin_Green_Si_solar_cell_review_history_1993.
pdf. Accessed on February 25, 2015.

Hamilton, James. 2015. "Careers in Solar Power." U.S.
Bureau of Labor Statistics. http://www.bls.gov/green/solar_
power/. Accessed on February 27, 2015.

Hanna, Harvey. 2010. "The Use of Magnifying Lenses in the
Classical World." Academia.edu. https://www.academia
.edu/467038/The_Use_of_Magnifying_Lenses_in_the_
Classical_World. Accessed on February 10, 2015.

Hermann, Allen. 2011. "Lecture 3: Types of Solar
Cells." http://www.ier.unam.mx/lifycs/ITaller2011/
PV-Tutorial2011/Lecture3-AH.pdf. Accessed on
February 27, 2015.

"History of Energy Consumption in the United States,
1775–2009." 2011. U.S. Energy Information
Administration. http://www.eia.gov/todayinenergy/detail
.cfm?id=10. Accessed on February 16, 2015.

"The History of Solar." 2015. Energy Efficiency and
Renewable Energy. https://www1.eere.energy.gov/solar/
pdfs/solar_timeline.pdf. Accessed on February 18, 2015.

"How Do Solar Panels Work?" 2015. The Power of Redarc
Solar. http://www.redarc.com.au/solar/about/solarpanels.
Accessed on July 22, 2015.

"How Solar Panels Are Made." 2015. Yingli Solar. http://
www.yinglisolar.com/us/solar-basics/#panel_section.
Accessed on July 22, 2015.

"Hybrid Energy." 2015. EcoPlanetEnergy. http://www
.ecoplanetenergy.com/all-about-eco-energy/overview/
hybrid/. Accessed on February 28, 2015.

"The Increasing Efficiency of Renewable Energy." 2013.
The Green Age. http://www.thegreenage.co.uk/
renewable-efficiency-price-drop-per-watt/. Accessed on
February 26, 2015.

Jayaram, V. 2015. "Hinduism—Gods and Goddesses in the Veda." http://www.hinduwebsite.com/hinduism/vedicgods .asp. Accessed on February 7, 2015.

Jordan, Borimir, and John Perlin. 1979–1980. "Solar Energy Use and Litigation in Ancient Times." *Solar Law Reporter.* 1(3): 583–594.

Koff, Stephen. 2005. "Was Jimmy Carter Right?" Resilience. http://www.resilience.org/stories/2005-10-12/ was-jimmy-carter-right. Accessed on February 25, 2015.

Kramer, Alexandra. 2014. "Solar Powered Planes, Cars and Boats Set New Records in Sustainable Transportation." Kenergy Solar. http://kenergysolar.com/blog/2014/5/8/ solar-powered-planes-cars-and-boats-set-new-records- in-sustainable-transportation. Accessed on March 1, 2015.

Krystek, Lee. 2011. "Archimedes and the Burning Mirror." The Museum of Unnatural History. http:// www.unmuseum.org/burning_mirror.htm. Accessed on February 10, 2015.

Layard, Austen Henry. 1853. *Discoveries in the Ruins of Nineveh and Babylon.* London: John Murray. https:// archive.org/details/discoveriesinru00layagoog. Accessed on February 10, 2015.

Lienhard, John H. 2013. "No. 2871: Solar Power in 1884." Engines of Our Ingenuity. http://www.uh.edu/engines/ epi2871.htm. Accessed on February 15, 2015.

Mints, Paula. 2013. "Solar PV Profit's Last Stand." Renewable Energy World.com. http://www.renewableenergyworld .com/rea/news/article/2013/03/solar-pv-profits-last-stand. Accessed on February 26, 2015.

"Monthly Energy Review. February 2015." 2015. Energy Information Administration. http://www.eia.gov/ totalenergy/data/monthly/pdf/mer.pdf. Accessed on February 26, 2015.

"Names of Gods and Goddesses." 2008. http://www
.lowchensaustralia.com/names/gods.htm. Accessed on
February 7, 2015.

Nersesian, Roy L. 2007. *Energy for the 21st Century:
A Comprehensive Guide to Conventional and Alternative
Sources.* Armonk, NY: M.E. Sharpe.

"The Nimrud Lens/The Layard Lens." 2015. British Museum.
http://www.britishmuseum.org/research/collection_online/
collection_object_details.aspx?objectId=369215&partId=1.
Accessed on February 10, 2015.

"19th Century Solar Energy Engines." 2012.
SolPower People. http://solpowerpeople.com/
solar-energy-engines-in-the-19th-century/. Accessed on
February 13, 2015.

"The 19th Century Solar Engines of Augustin Mouchot,
Abel Pifre, and John Ericsson." 2012. Land Art Generator
Initiative. http://landartgenerator.org/blagi/archives/2004.
Accessed on February 13, 2015.

Nixon, Richard M. 1973. "Address to the Nation about
National Energy Policy." The American Presidency Project.
http://www.presidency.ucsb.edu/ws/?pid=4051. Accessed
on February 19, 2015.

Noth, André. 1980. "Design of Solar Powered Airplanes for
Continuous Flight." Doctoral Dissertation, ETH Zürich.
http://www.sky-sailor.ethz.ch/docs/Thesis_Noth_2008.pdf.
Accessed on March 1, 2015.

Obama, Barack. 2009. "Address to Joint Session of
Congress." http://www.whitehouse.gov/the_press_office/
Remarks-of-President-Barack-Obama-Address-to-Joint-
Session-of-Congress/. Accessed on February 5, 2015.

"October 28, 1980 Debate Transcript." 1980. Commission
on Presidential Debates. http://www.debates.org/index.
php?page=october-28-1980-debate-transcript. Accessed on
February 25, 2015.

"Oil Embargo, 1973–1974." 2013. U.S. Department of State Office of the Historian. https://history.state .gov/milestones/1969-1976/oil-embargo. Accessed on February 19, 2015.

O'Kelly, Michael J. 1998. *Newgrange: Archaeology, Art and Legend.* London: Thames and Hudson.

Olcott, William Tyler. 2008. *Sun Lore of All Ages.* Charleston, SC: Bibliobazaar. Reprint of the original 1914 edition. http://www.sacred-texts.com/astro/slaa/slaa08.htm#fn_43. Accessed on February 7, 2015.

Parry, Sam. 2012. "Reagan's Road to Climate Perdition." Consortiumnews.com. http://consortiumnews .com/2012/01/29/reagans road to climate perdition/. Accessed on February 25, 2015.

Perlin, John. 1986. "Ancient Greek Solar Architecture: Lessons for Today's Architect." In E. Bilgen, ed. 1986. *INTERSOL 85: Proceedings of the Ninth Biennial Congress of the International Solar Energy Society: Montreal, Canada, 23–29 June 1985.* New York: Pergamon Press.

Perlin, John. 2004. "The Silicon Solar Cell Turns 50." National Renewable Energy Laboratory. http://www.nrel .gov/education/pdfs/educational_resources/high_school/ solar_cell_history.pdf. Accessed on February 18, 2015.

Perlin, John. 2005. "Solar Evolution." http://www .californiasolarcenter.org/history.html. Accessed on February 6, 2015.

Perlin, John. 2015. "The Story of Vanguard." UCSB Experimental Cosmology Group. http://www.deepspace .ucsb.edu/outreach/the-space-race/the-story-of-vanguard. Accessed on February 18, 2015.

"The Photoelectric and Photovoltaic Effects." 2015. Sargosis Solar & Electric. http://sargosis .com/articles/science/how-pv-modules-work/

the-photoelectric-and-photovoltaic-effects/. Accessed on February 17, 2015.

"Photovoltaics: Solar Cells 101." 2015. Alpenglow Solar. http://alpenglowsolar.com/solar-cells-101.php. Accessed on February 27, 2015.

"Photovoltaics Report." Fraunhofer Institute for Solar Energy Systems ISE. http://www.ise.fraunhofer.de/de/downloads/pdf-files/aktuelles/photovoltaics-report-in-englischer-sprache.pdf. Accessed on February 28, 2015.

"Public Law 93–409." 1974. http://www.gpo.gov/fdsys/pkg/STATUTE-88/pdf/STATUTE-88-Pg1069.pdf. Accessed on February 19, 2015.

Rajagopalan, Rajeswari Pillai, and Rahul Prakash. 2011. "China Walks the US-India Space Solar Power Dream." Observer Research Foundation. http://www.orfonline.org/cms/sites/orfonline/modules/analysis/AnalysisDetail.html?cmaid=25917&mmacmaid=25918. Accessed on March 1, 2015.

"R&D Priorities within the Department of Energy." 2015. American Physical Society. http://www.aps.org/policy/reports/popareports/energy/doe.cfm. Accessed on February 25, 2015.

"Renewable Energy Fact Sheet: Solar Cells." 2013. United States Environmental Protection Agency. http://water.epa.gov/scitech/wastetech/upload/Solar-Cells.pdf. Accessed on February 26, 2015.

Richmond, Michael. 2015. "Einstein and the Photoelectric Effect." http://spiff.rit.edu/classes/phys314/lectures/photoe/photoe.html. Accessed on February 17, 2015.

Robinson, David M., and J. Walter Graham. 1938. *Excavations at Olynthus, Part VIII: The Hellenic House; a Study of the Houses Found at Olynthus with a Detailed Account of Those Excavated in 1931 and 1934.* Baltimore: John Hopkins Press.

Rosenblum, Louis, et al. 1979. "Photovoltaic Power Systems for Rural Areas of Developing Countries." NASA Center for Aerospace Information. http://ntrs.nasa.gov/archive/ nasa/casi.ntrs.nasa.gov/19790007240.pdf. Accessed on February 19, 2015.

"Russell Ohl Accidentally Discovers the Silicon P-N Junction." 1999. PBS.org. http://www.pbs.org/ transistor/science/events/pnjunc.html. Accessed on February 17, 2015.

Siemens, Werner. 1885. "On the Electromotive Action of Illuminated Selenium, Discovered by Mr. Fritts of New York." *Van Nostrand's Engineering Magazine*. 32: 514–516. https://books.google.com/books?id=0CdCAQAAIAAJ&p g=PA514&lpg=PA514&dq=%22on+the+electromotive+a ction+of+illuminated+selenium%22+van+nostrand%27s +engineering+magazine&source=bl&ots=UxIbCN6HbW &sig=0Xq-XenMrtKx3mGlikvqEHw8tyY&hl=en&sa=X &ei=_IriVL3gHcfXoATJmYLAAw&ved=0CCMQ6AEw AQ#v=onepage&q=%22on%20the%20electromotive%20 action%20of%20illuminated%20selenium%22%20 van%20nostrand's%20engineering%20magazine&f=false. Accessed on February 16, 2015.

Simonin, Louis Laurent. 1876. "Industrial Applications of Solar Heat." *Popular Science Monthly*. 9: 550–560. (Translated by J. Fitzgerald). http://en.wikisource.org/wiki/ Popular_Science_Monthly/Volume_9/September_1876/ Industrial_Applications_of_Solar_Heat. Accessed on February 13, 2015. [This article is a translation of the original French article: "L'Emploi Industriel la Chaleur Solaire." May 1876. *Revue Des Deux Mondes*. http://www .revuedesdeuxmondes.fr/archive/article.php?code=63413. Accessed on February 13, 2015.]

Smith, Charles. 1995. Revisiting Solar Power's Past. Technology Review. http://solarenergy.com/power- panels/history-solar-energy. Accessed on February 16, 2015.

Smith, Willoughby. 1873. "The Action of Light on Selenium." *Journal of the Society of Telegraph Engineers*. 2: 31–33.

"Solar Battery Turns Sunshine into Usable Electric Power." 1954. Popular Mechanics. June 1954: 71. Jflgs://books. google.com/books?id=1t4DAAAAMBAJ&pg=PA71&lp g=PA71&dq=%22compares+favorably+with+the+efficie ncy+of+steam+and+gasoline+engines%22&source=bl& ots=VGDB6djBIw&sig=RVaQ_vf5TTmLT87ipNFyjAL JFlg&hl=en&sa=X&ei=6MPkVKiRD8yuogSXmYGgC Q&ved=0CCQQ6AEwAQ#v=onepage&q=%22compa res%20favorably%20with%20the%20efficiency%20of%20- steam%20and%20gasoline%20engines%22&f=false. Accessed on February 18, 2015.

"Solar Electricity Costs." 2015. Solar Cell Central. http:// solarcellcentral.com/cost_page.html. Accessed on February 26, 2015.

"Solar Energy 101: Introduction to Solar Energy." 2015. Dow Corning. http://www.dowcorning.com/content /solar/solarworld/solar101.aspx. Accessed on July 22, 2015.

"Solar Energy Converting Apparatus, US2780765A." 1957. http://www.google.com/patents/US2780765. Accessed on February 18, 2015.

"Solar Energy Shows Its Paces." 1983. New Scientist. November 10, 1983: 404. https://books.google.com/boo ks?id=YFjTMckHfuwC&pg=PA404&lpg=PA404&dq= hesperia+arco+solar+1982&source=bl&ots=HP6x8CpSn K&sig=5vW4Ip2YdNI6u6PiFqhqx0GnXSE&hl=en&sa =X&ei=fm_mVMnzB4u7ogTFtoLYDg&ved=0CDUQ6 AEwAw#v=onepage&q=hesperia%20arco%20solar%20 1982&f=false. Accessed on February 19, 2015.

Solar Impulse Beta. 2015. http://www.solarimpulse.com/. Accessed on March 1, 2015.

"Solar Power Products." 2015. Solar Switch. http://www
.solarswitchaustralia.com.au/solar-power-products.
Accessed on February 28, 2015.

"Solar Stats." 2012. Home Energy. http://homeenergyllc.com/
archives/1372. Accessed on February 26, 2015.

"Space Solar Power." 2014. National Space Society. http://
www.nss.org/settlement/ssp/. Accessed on March 1, 2015.

Taylor, Ken. 2012. *Celestial Geometry: Understanding the
Astronomical Meanings of Ancient Sites*. London: Watkins
Publishing.

Thiele, Timothy. 2015. "Top 10 Solar Energy Uses." About.
home. http://electrical.about.com/od/appliances/tp/
Top-10-Solar-Energy-Uses.htm. Accessed on March 1, 2015.

"This Month in Physics History: April 25, 1954: Bell Labs
Demonstrates the First Practical Silicon Solar Cell."
2009. APS Physics. http://www.aps.org/publications/
apsnews/200904/physicshistory.cfm. Accessed on
February 17, 2015.

"Timeline of Solar Cells." 2015. USA Alternative Energy
NOW! http://usaalternativeenergynow.blogspot
.com/2011/10/timeline-of-solar-cells.html. Accessed on
February 17, 2015.

"2010 Solar Technologies Market Report." 2011. U.S.
Department of Energy. Energy Efficiency and Renewable
Energy. http://www.nrel.gov/docs/fy12osti/51847.pdf.
Accessed on February 26, 2015.

"2013 Renewable Energy Data Book." 2014. U.S.
Department of Energy. Energy Efficiency& Renewable
Energy. http://www.nrel.gov/docs/fy15osti/62580.pdf.
Accessed on February 26, 2015.

"Types of Solar Energy." 2015. Solar Energy at Home. http://
www.solar-energy-at-home.com/types-of-solar-energy.html.
Accessed on February 27, 2015.

"Vanguard 1." 2014. National Aeronautics and Space Administration. http://nssdc.gsfc.nasa.gov/nmc/spacecraftDisplay.do?id=1958-002B. Accessed on February 18, 2015.

Vendel, Ottar. 2015. "115 Egyptian Gods." http://www.nemo.nu/ibisportal/0egyptintro/1egypt/. Accessed on February 7, 2015.

"What Is Solar Energy?" 2014. The Eco Ambassador. http://www.theecoambassador.com/solarenergy.html. Accessed on February 27, 2015.

Wihbey, John. 2008. "Jimmy Carter's Solar Panels: A Lost History That Haunts Today." Yale Climate Connections. http://www.yaleclimateconnections.org/2008/11/jimmy-carters-solar-panels/. Accessed on February 25, 2015.

Wik, Stephen R. 2013. *How the Ray Gun Got Its Zap: Odd Excursions into Optics*. Oxford, UK: Oxford University Press.

"William Cobb Demonstrates First Solar-Powered Car." 2014. This Day in History. http://www.history.com/this-day-in-history/william-cobb-demonstrates-first-solar-powered-car. Accessed on March 1, 2015.

Willsie, H. E. 1909. "Experiments in the Development of Power from the Sun's Heat." *Engineering News*. May 13: 511–514.

Wilson, Gregory M. 2013. "Building on 35 Years of Progress—The Next 10 Years of Photovoltaic Research at NREL." National Renewable Energy Laboratory. https://www.purdue.edu/discoverypark/energy/assets/pdfs/pdf/Pioneer%20in%20Energy%20Greg%20Wilson%20Presentation%207.17.13.pdf. Accessed on February 25, 2015.

Wood, Daniel. 2014. "Space-Based Solar Power." Energy.gov. http://energy.gov/articles/space-based-solar-power. Accessed on March 1, 2015.

2   Problems, Controversies, and Solutions

"We also see an exponential progression in the use of solar energy. It is doubling now every two years. Doubling every two years means multiplying by 1,000 in 20 years. At that rate we'll meet 100 percent of our energy needs in 20 years."
(Futurist Ray Kurzweil, as quoted in Lloyd 2008)

Solar power will be the single largest source of electricity generation by the mid-point of the century.
(Tesla CEO and SolarCity chairman Elon Musk, as quoted in Braun 2014)

Comments such as these by proponents and supporters of solar energy sound very optimistic. They suggest that solar energy will play a significant role in meeting the energy needs of the United States and many other nations of the world both in the near future and in the long term. Such views may seem overly optimistic and unrealistic to some observers. Any graph summarizing the production and consumption of solar energy in the past is essentially a flat line, running very near the zero coordinate on the y-axis (for example, see "Where Did All the Solar Go?" 2012), and some projections for the future do not hold much promise for a significant change in that trend. For example, the U.S. Energy Information Administration (EIA) predicts that the increase in solar energy production over the next 20 years is likely to be relatively modest, with a curve from 2015 to 2035 looking not

A worker checks solar cells at the ErSol Solar Energy factory in Erfurt, Germany. (AP Photo/Jens Meyer)

so very different from that for 2005 to 2015, that is, a straight line along the horizontal axis of the graph ("Global Installed Power Generation Capacity by Renewable Source" 2011).

Solar Energy Worldwide

One question worth asking is how typical is the experience of the United States in developing solar energy compared to other parts of the world. At the end of 2013, the most recent year for which data are available, the United States ranked fifth in the world, with a cumulative installed photovoltaic (PV) capacity of 12.1 gigawatts. A gigawatt (GW) is equivalent to one billion watts of power. The world's leader in solar power on this measure was Germany, with a cumulative installed capacity of 35.9 gigawatts, followed by China (19.9 GW), Italy (17.6 GW), Japan (13.6 GW), Spain (5.6 GW), France (4.6 GW), the United Kingdom (3.3 GW), Australia (3 GW), and Belgium (3 GW). All other countries combined had an installed capacity of 20.2 GW ("Global Cumulative Solar PV Capacity at the End of 2013, by Country [in Gigawatts]" 2015).

Global trends in the production of solar photovoltaics roughly correspond to that in the United States, with a relatively flat curve about to the late 2000s, followed by an accelerating development of solar PVs after 2010, as shown in Table 2.1. Growth has not been the same for various countries, however. As Table 2.2 shows, China has had by far the greatest development of solar PV technology of any nation in the world, followed by Japan and the United States. It is interesting to note that nearly all of the world's developing countries, most of them in regions where solar energy is abundant (such as the Middle East, Africa, South America, and most of East Asia), have essentially no solar PV facilities. The total installed solar PV capacity in all of the Middle East in 2012, for example, was only 0.4 GW, all of which was located in two countries, Saudi Arabia and the United Arab Emirates. The total installed PV solar capacity in Africa in 2012, similarly, was only 0.3 GW, nearly 70 percent of which was located in Egypt ("International Energy Statistics" 2015).

Table 2.1   Global Trends in the Production of Photovoltaics
(in Gigawatts)

Year	Total Capacity
2004	3.7
2005	5.1
2006	7
2007	9
2008	16
2009	23
2010	40
2011	70
2012	100
2013	139

Source: REN21. 2014. *Renewables 2014 Global Status Report*. (Paris: REN21 Secretariat) ISBN 978–3–9815934–2–6. Figure 12, 49. http://www.ren21.net/portals/0/documents/resources/gsr/2014/gsr2014_full%20report_low%20res .pdf. Accessed on March 2, 2015. Used by permission of REN21.

Table 2.2   Increase in Solar PV Capacity in 2013 (in Gigawatts)

Country	Increase in Capacity (GW)
China	12.9
Japan	6.9
United States	4.8
Germany	3.3
Italy	1.5
United Kingdom	1.5
Australia	0.8
France	0.6
Spain	0.2
Belgium	0.2

Source: REN21. 2014. *Renewables 2014 Global Status Report*. (Paris: REN21 Secretariat) ISBN 978–3–9815934–2–6. Figure 13, 49. http://www.ren21.net/portals/0/documents/resources/gsr/2014/gsr2014_full%20report_low%20res .pdf. Accessed on March 2, 2015. Used by permission of REN21.

Concentrating solar power (CSP) technology has yet to become very popular throughout the world. As of 2014, only 19 countries have operating CSP plants, with one of them, Spain, accounting for about 72 percent of all the solar energy harvested by this technology. The United States accounts for a quarter of all CSP capacity, and the remaining seven nations (Algeria, Australia, Egypt, India, Italy, Morocco, and Thailand) account for the remaining CSP capacity ("Renewables" 2014, 51; "Spain Accounts for 72% of the World's Concentrating Solar Power" 2012).

Solar heating and cooling has also been showing increases in capacity over the past decade similar to those for solar PV and CSP technologies. Worldwide capacity more than tripled in the decade from 2003 to 2013, to its 2013 value of 326 GW of power. Nearly two-thirds of that capacity was located in China (64 percent), followed by the United States (5.8 percent), Germany (4.2 percent), Turkey (3.9 percent), Brazil (2.1 percent), and India (1.6 percent) ("Renewables" 2014, 54).

These data provide a glimpse of the role that solar energy is playing in nations around the world that is not so very different from that in the United States. They suggest that nations are on the verge of making increasingly greater commitments to the use of solar PV, CSP, and thermal solar heating and cooling technologies to help meet their energy needs. In order to see continued progress, however, developers of solar energy will have to solve a number of technical, social, economic, environmental, political, and other problems.

Problems Facing the Solar Industry

One of the most obvious problems with which the solar industry must deal is the intermittency of solar energy. The Sun does not always shine with the same intensity at all times of the day in all places at various times of the week and year. For example, Table 2.3 shows the average insolation reaching various points on Earth's surface at different times of the year. The term *insolation* refers to the amount of solar radiation reaching a given area on Earth's surface. The data shown in this table can be more easily understood by referring to a graph of the two variables like the one found at Earth Observatory, http://earthob servatory.nasa.gov/Features/EnergyBalance/page3.php.

Table 2.3 Monthly Latitude Insolation (Watts per Square Meter)

Lat	Jan	Feb	Mar	Apr	May	Jun	Jul	Aug	Sep	Oct	Nov	Dec	Annual
-90	496	305	58	0	0	0	0	0	10	214	444	552	172.8
-80	489	301	101	6	0	0	0	0	51	221	437	543	178.6
-70	468	323	170	54	4	0	0	29	121	265	422	519	197.4
-60	472	367	237	122	50	25	37	91	190	318	440	505	237.2
-50	488	407	297	190	114	83	98	158	253	365	463	512	285.2
-40	498	437	349	254	182	149	164	223	310	403	478	514	329.6
-30	497	456	390	311	247	216	230	283	357	429	483	507	366.7
-20	484	462	419	360	307	280	292	336	393	443	475	487	394.5
-10	458	455	435	398	359	338	347	380	418	444	454	456	411.6
0	421	435	439	425	402	388	393	412	429	432	421	413	417.4
10	372	402	429	439	435	428	429	433	428	407	377	360	411.7
20	314	358	406	440	455	458	454	442	414	370	323	299	394.6
30	249	303	371	429	464	476	468	438	388	322	261	231	366.9
40	179	241	324	405	461	483	470	422	349	265	193	160	329.9
50	108	173	268	370	447	481	462	395	301	201	123	89	285.5
60	42	103	204	325	427	475	448	359	243	132	56	27	237.5
70	1	36	135	275	413	488	447	320	179	63	6	0	197.8
80	0	1	63	237	428	512	468	305	112	10	0	0	179.0
90	0	0	20	235	435	520	475	310	72	0	0	0	173.2

Source: Monthly Latitude Insolation. 2013. National Aeronautics and Space Administration. http://data.giss.nasa.gov/ar5/srmonlat.html; http://data.giss.nasa.gov/cgi-bin/ar5/srmonlat.cgi. Accessed on March 5, 2015.

Notice that the Earth receives a relatively high amount of solar energy year-round at the equator. At higher latitudes, however, Earth's surface receives less—sometimes much less—solar radiation during the winter months in the Northern Hemisphere and more radiation during the summer months (with the reverse situation in the Southern Hemisphere). Thus, solar power plants would, in general, be expected to work quite well in the summer months in the middle latitudes but not so well in the higher latitudes, a fairly obvious statement of one's everyday experience. Because solar radiation is so variable throughout the year and at various latitudes, it is sometimes said to be a *variable generation* source of energy. Wind is another example of a renewable energy source classified as a variable generation source because winds blow with variable speeds at various times and various places.

A critical resource for companies planning to build solar power plants, then, would be some type of resource that shows insolation patterns over an area at various times of the year. Over time, the U.S. National Renewable Energy Laboratory (NREL) has developed a collection of such maps for use by solar power developers and other interested parties. These maps show insolation patterns in the United States and around the world for all seasons of the year based on data collected from 1998 to 2009. Two sets of maps are available, those that show insolation patterns most favorable for solar PV systems and those most favorable for CSP systems. These maps are available at the NREL website at "Dynamic Maps, GIS, & Analysis Tools," http://www.nrel.gov/gis/solar.html. Similar insolation maps have been developed for various parts of the world (such as Europe or Africa), as well as for the complete globe (see, for example, "Annual Solar Irradiance, Intermittency, and Annual Variations" at http://www.greenrhinoenergy.com/solar/radiation/empiricalevidence.php).

Latitude and time of the year are, of course, only two factors affecting solar insolation at a particular location. The NREL maps described earlier are a very useful tool in the search for solar power sites, but only a rather gross tool. Another important factor needed for refining a site search is weather patterns. The cloudier the weather in a region and the longer those clouds persist, the less insolation for the area. Again, this fact simply reflects everyday experience that, all other factors being equal, a solar power plant is likely to be able to produce more electricity in Phoenix, Arizona, than in Seattle, Washington. An important feature of the planning for possible solar power plants, then, is research on the amount of solar radiation received at the proposed site for the plant and the general pattern over time for the radiation. An example of the type of data needed for such an exercise can be found "The Sun's Energy," at http://www.powerfromthesun.net/Book/chapter02/chapter02.html, Figure 2.14 (see also "Texas Renewable Energy Resource Assessment 2008," at http://www.seco.cpa.state.tx.us/publications/renewenergy/solarenergy.php, Exhibit 3–12.)

Locating a solar plant in a region with maximum sunshine throughout the year is, therefore, a basic technical problem that designers have to solve in order to have an efficient facility. They have developed a number of possible solutions for this problem. For example, one way of dealing with insolation intermittency is to combine the output of a number of solar plants on the same grid. The efficiency in each individual plant, then, can be compensated for by the efficiencies of other solar plants on the grid. An illustration as to how this plan can be achieved is shown in the graph at "Variability of Renewable Energy Sources," at http://www.nrel.gov/electricity/transmission/variability.html. (The term *electrical grid* refers to all the providers and consumers of electrical power along with the system of wires that connects them to each other, operated by one or more control centers.)

Storage of Solar Energy

One of the most basic problems posed by the use of solar energy and wind energy, then, is to find a way to make sure the systems can produce electricity all the time, day or night, any time of the year. This restriction is not a problem, of course, for conventional fossil fuel plants, where coal, oil, or natural gas can be provided to a furnace at any time, ensuring that the plant produces electricity continuously. The same is true for a nuclear power plant. But designers of solar (and wind) facilities have to find ways to make sure the plant continues to produce electricity when the Sun doesn't shine or the wind doesn't blow. This problem is one of *storage*, that is, collecting and saving the solar (or wind) energy produced at optimal times so that it will be available for use in generating electricity at suboptimal times. The system for storing solar energy differs from facility to facility, depending on whether the facility is designed for the generation of electricity or solar heating applications. The problem also differs substantially for small residential solar systems and larger commercial systems.

Residential PV Solar

The simplest method of storing solar energy in a residential solar PV system is by using excess solar energy to charge a battery or bank of batteries. During sunny periods, solar radiation is captured by a traditional solar PV system and transferred to the electrical system in a home. The electricity is used for conventional purposes, such as lighting the house and operating appliances. Any excess electricity not used for these purposes is used to charge a battery or set of batteries. During periods of low solar radiation, then, the batteries can be used to run all the electrical systems in the structure. Such systems are relatively simple in a residential home, but can become more complicated in larger office buildings or industrial plants. The larger the facility, the more batteries that may need to be charged.

The most serious problem involved in the use of batteries for the storage of solar energy is the cost of the batteries needed for storage. Historically, such systems have relied on familiar batteries, such as lead-acid, nickel cadmium (NiCad), lithium-ion, and sodium–sulfur batteries. But until recently, the cost of these batteries has been too high to permit their use in most residential settings. For example, one recent study found that the addition of battery storage to a residential solar PV system resulted in an increase in overall cost for the system of between 10 and 40 percent (DNV KEMA 2013, 26).

For this reason, a primary goal of many solar researchers and battery specialists today is to find ways of designing and making less expensive, more efficient solar storage batteries. That effort appears to have had some success as the cost of solar PV storage system has continually decreased over the past decade at a rate of about 11 percent per year (DNV KEMA 2013, 27). This trend has encouraged many proponents of battery storage as a promising addition to any residential solar PV system. Still, over the long term, batteries have provided only a minimal fraction of the storage capacity for solar PV power in residences. In the state of California, for example, between 2007 and 2013 the number of residences with battery storage ranged from 10 to 75, while those without battery storage ranged from 3,420 to 28,301, the largest differential occurring in 2012, when 29 homes had battery storage and 28,301 homes had solar PV systems without battery storage (DNV KEMA 2013, 25; for an excellent review of the current status of battery storage for residential solar PV systems, see Hoppmann et al. 2014; for a similar review of large-scale commercial solar PV systems, see Poullikkas 2013).

Because of the high cost of batteries, by far the preferred method of storage for solar PV systems is simply to "dump" it into the electrical grid. That is, any excess electricity produced by a residential system that cannot be used for normal household operations or for charging storage batteries is sent to the electrical grid. (A diagram of a system that contains both battery storage and connection to the grid can be found at "High Capacity 30 KWh Solar Lithium Batteries," http://www .hecobattery.com/china-high_capacity_30kwh_solar_lith-ium_batteries_48v_800ah_lifepo4_battery_bank_for_solar_ energy_storage-1303686.html.)

In such a case, the producer of that excess electricity may (or may not) receive a payment from the utility that operates the grid for the electricity it receives. This system is called *net metering* and is now in use in many parts of the United States and other parts of the world.

Rules and regulations controlling the use of net metering vary substantially from state to state and involve the amount of electricity that can be dumped, how much the utility is required to pay for the electricity (if anything), whether payments can be rolled over from month to month, and other factors. Each year, the Interstate Renewable Energy Council (IREC) and Vote Solar publish a report titled "Freeing the Grid" that provides statistics on net metering in the United States. The report assigns grades of A, B, C, D, F, or NA to each state's net metering policies. The grade is based on the type of rebate paid to customers for the dumping of excess electricity and the extent to which state policies encourage the practice of net metering. In its most recent report, IREC and Vote Solar assigned their highest scores to Colorado, with a score of 25 out of a possible 26.5, Delaware (23.5), Pennsylvania (23), California (22.5), Maryland (22.5), New Jersey (22), and Vermont (20.5). Lowest scores went to Georgia (0.5), Oklahoma (3), North Dakota (4), South Carolina (4.5), Virginia (5), Wisconsin (5.5), and Montana (7). Six states (Alabama, Idaho, Mississippi, South Dakota, Tennessee, and Texas) had no policies for net metering and so received no scores for the policy ("Freeing the Grid 2015: Best Practices in State Net Metering Policies and Interconnection Procedures" 2015, 102, 104).

Enthusiasm for net metering is not universal. Electric utilities in particular have expressed concern about the growth of stand-alone, usually rooftop, solar panel systems from which excess electricity flows into utility grids. Utilities claim that their grids were not built to *receive* electrical current but to *distribute* it. The wildly fluctuating current that comes from individual homes can cause all manner of problems for the grids, then, including voltage fluctuations that can damage electrical circuits leading to brownouts and blackouts. In addition, of course, net metering tends to reduce the income and profit realized by a public utility (Cardwell 2015, A4).

Commercial Solar Storage

Batteries can also be used in storage systems for very large, commercial solar PV generation facilities, almost always in connection with other storage technologies. For a rendering of the type of facility that might be used for such a purpose, see "Energy Storage Industry Grows to Integrate Wind, Solar," at http://www.renewableenergyworld.com/rea/news/article/2011/08/energy-storage-industry-grows-to-integrate-wind-solar. As with residential systems, the primary challenge for commercial solar storage systems is to produce batteries that can do the job required at a reasonable economic cost, a challenge that has yet to be met. A number of research projects are underway, however, to make more efficient batteries and integrate them into storage systems for commercial solar PV power systems (see, for example, AzRISE Energy Storage Initiative 2012).

Given the cost of battery storage, commercial solar storage systems, like their residential counterparts, almost always rely on a different type of storage, pumped hydro. The term *pumped hydro* refers to a system in which water is pumped for a lower water reservoir through a system of pipes into a higher water reservoir. In a renewable energy storage system, the electricity required to pump water uphill comes from (usually) wind or solar facilities that produce excess electricity under favorable wind or solar conditions during the day. Under less favorable conditions, typically at night, the water is allowed to flow downhill from the upper reservoir to the lower reservoir, where it is used to drive a turbine and generator ("Challenges and Opportunities for New Pumped Storage Development" 2012; for a diagram of a solar power pumped hydro storage system, see "Storing Energy," http://www.creighton.edu/green/energytutorials/formsofenergy/storingenergy/).

As of 2015, pumped hydro is by far the most common method for storing energy produced by renewable energy sources such as wind and solar. It is responsible for the production of 247,010 megawatt-hours (MWh) of electrical power from storage in the United States, compared to 310 MWh from batteries, 6,050 MWh from compressed air, and 1,621 MWh from thermal storage systems ("Market Analysis: Current US Energy Storage Market" 2014, 4). The process also accounts for about 99 percent of solar energy storage worldwide (Matteson 2015). The major drawback for pumped hydro systems is usually the availability of elevated regions on which upper reservoirs can be built. They are not very effective, for example, on flat plains in sunny states of the U.S. Southwest.

The technologies required for storage of energy produced by CSP are different from those for solar photovoltaics. The general approach is to collect any heat produced by CSP processes that is not used immediately and find a way to retain that heat in some type of medium so that it can be used at a later time. One type of technology involves using the CSP-generated heat to warm up a material that has a high heat capacity, that is, a material that requires a lot of energy in order to raise its temperature by 1° Celsius. The material should also have other physical properties, such as a high density. A high density allows a designer to pack a large quantity of the material into a relatively small space. Some materials that have been studied for possible use as a storage material for CSP systems are reinforced concrete, cast iron, silica fire blocks, liquid sodium, silicone oils, and a variety of inorganic compounds, such as lithium bromide, sodium chloride, potassium hydroxide, and sodium nitrate (Kuravi et al. 2013).

An alternative approach for the storage of CSP-generated heat involves the use of a reversible chemical reaction. For example, the decomposition of ammonia (NH_3) is an endothermic reaction (one that requires the input of heat):

$$NH_3 + heat \rightarrow \tfrac{1}{2}N_2 + \tfrac{3}{2}H_2$$

The amount of heat needed to bring about this reaction is 67 kJ/mol (kilojoules per mole). Thus, if a hot fluid from a CSP system is fed into a tank of ammonia gas, the gas will decompose into nitrogen and hydrogen gases. The heat from the CSP system gets stored in the two elements, nitrogen and hydrogen. At some later time, the two gases can be brought into contact to bring about the reverse of the reaction shown earlier, resulting in the formation of ammonia gas:

$$\tfrac{1}{2}N_2 + \tfrac{3}{2}H_2 \rightarrow NH_3 + heat$$

The formation of ammonia from its elements is, thus, accompanied by the release of the same amount of energy absorbed during its decomposition. That released energy can then be transmitted to some end point for use, just as had the original CSP-generated energy ("Solar Resources" 2014; "Ammonia–Based Energy Storage" 2014).

Environmental Issues

To acknowledge that solar energy may pose problems for the environment can be a somewhat startling claim. After all, one of the most important reasons for supporting and promoting solar energy is that it is generally thought to be much less harmful to the natural environment than traditional fossil fuel resources, such as coal, oil, and natural gas, and less risky by far than nuclear energy. And, indeed, a number of researchers have attempted to quantify the threat posed to human health as a result of using various forms of energy source. Such efforts are subject to some questions because of the difficulty in associating the burning of coal or operating a nuclear power plant, for example, with damage to human health. But Table 2.4 provides the results of such calculations that probably reflect the views of many specialists in the field (for another calculation of this kind, also see Wang 2011).

Table 2.4   Death Rate Due to Various Types of Energy Use (in Deaths per
           Trillion Kilowatt Hours)

Energy Source	Death Rate
Coal (world average)	170,000
Coal (China)	280,000
Coal (United States)	15,000
Oil	36,000
Natural gas	4,000
Biofuels	24,000
Solar	440
Wind	150
Hydroelectric	1,400
Nuclear	90

Source: Conca, James. 2012. "How Deadly Is Your Kilowatt? We Rank the Killer
Energy Sources." *Forbes*. http://www.forbes.com/sites/jamesconca/2012/06/10/
energys-deathprint-a-price-always-paid/. Accessed on March 5, 2015.

Solar Cell Manufacture

The question as to the extent to which solar PV manufacture
and use cause environmental problems is one about which
there is still some disagreement. Some individuals and agencies
have pointed to a number of points in the PV cell life cycle at
which such problems may arise. For example, a 2009 white
paper by the Silicon Valley Toxics Coalition (SVTC) listed a
number of steps in the conversion of silicon dioxide and other
raw materials to finished solar cell, cell use, and disposition
of the cell at which environmental risks may occur. The type
of potential hazards depends very much on the type of cell
being manufactured (crystalline silicon [c-Si], amorphous silicon [a-Si], cadmium telluride [CdTe], copper indium selenide
[CIS], etc.), so the following discussion is only a sample of the
types of concerns expressed in the SVTC white paper. Among
the instances mentioned in that paper that may involve environmental hazards were the following (Mulvaney et al. 2009):

- Silicon dioxide mining, the first step in the solar PV cycle,
  may pose health threats to humans because of the release of

silicon dioxide particles to the air during the mining process. Silica (silicon dioxide) dust has long been known to cause a type of respiratory disorder known as silicosis that may cause such short-term problems as shortness of breath, coughing, and general weakness. Over time, these symptoms may become worse and ultimately lead to respiratory failure ("Symptoms, Diagnosis and Treatment" 2015).

- The manufacture of ultrapure silicon from raw silicon dioxide involves the use of a gas known as silane (SiH_4), a very hazardous substance. Silane is very unstable, with a tendency to explode unexpectedly. It may, therefore, pose a hazard to workers and also to the surrounding community (Ngai 2008).

- Individual silicon wafers used for the production of solar PV cells are made by sawing off very small segments of a pure silicon crystal. In the process, silicon dust (called *kerf*) is produced, posing potential silicosis risk for workers involved in the procedure. Although this health hazard is well known, some fraction of workers may still be at risk for silicosis-type disorders from exposure to kerf (Yassin, Yebesi, and Tingle 2005).

- A number of toxic substances are used or produced during the doping of silicon wafers in the process of making solar PV cells. (Recall that doping is the process of adding small amounts of impurities to a silicon wafer to improve its semiconducting characteristics. Common doping agents include antimony, arsenic, boron, and phosphorus.) Among the most widely used doping agents are the gases arsine (AsH_3) and phosphine (PH_3), both highly toxic. In one major report on the health hazards of the manufacture of solar PV cells, arsine and phosphine were said to pose "[t]he greatest environmental risk . . . during the manufacturing process" ("Potential Health and Environmental Impacts Associated With the Manufacture and Use of Photovoltaic Cells 2004," v; for health and fire risks associated with arsine and phosphine, see "Arsine (SA): Systemic Agent" 2014 and "Phosphine: Lung Damage Agent" 2014).

- A number of different caustic and otherwise harmful chemicals are used at various stages of the manufacturing process to clean silicon chips and the vessels used in their processing, such as sodium and potassium hydroxide; hydrochloric, sulfuric, and nitric acids; hydrogen fluoride; and isopropyl alcohol. One cleansing agent of particular concern is the gas sulfur hexafluoride (SF_6), which is now recognized by the Intergovernmental Panel on Climate Change as the most potent greenhouse gas ("2.10.2 Direct Global Warming Potentials 2007"; Mulvaney 2013).

- Other types of solar PV cells have other or additional potential environmental hazards. For example, cadmium-tellurium (CdTe) cells carry the additional burden that one of the basic materials used in the cell—cadmium metal—is thought to be a human carcinogen, with additional hepatic (liver), hematological (blood), and immunological effects on humans. Government agencies have established safe limits for exposure to cadmium at 0.04 mg/L (milligrams per liter) in drinking water and 5 μg/m³ (micrograms per cubic meter) in air ("Toxicological Profile for Cadmium" 2012, 8).

- Selenium hydride (SeH_2), a gas used to lay down the selenium layer in a copper indium selenide (CIS) or copper indium gallium selenide solar cell, has a variety of toxic properties. A number of studies have confirmed that the gas may pose a range of hazards to human health, including a variety of respiratory, gastrointestinal, and cardiovascular effects; conditions known as "blind staggers" and "alkali disease"; and possible carcinogenic and reproductive disorders ("Selenium Compounds" 2013; for additional information on this topic, see also Alsema 1996; Fthenakis and Bowerman 2015).

Solar Cell End-of-Life

The problems posed by toxic and hazardous chemicals during the production of solar PV cells do not disappear after the manufacturing process has been completed. Indeed, the SVTC report on the hazards associated with solar PV cell production points out that those problems (known as *end-of-life* problems) are likely to become an issue once more when cells have reached the end of their useful life, after about 25 years according to the author of the report (Mulvaney 2009, 19). At that point, cells are likely to end up in solid waste disposal sites, such as landfills, where they will bring with them a long-term legacy of toxic and hazardous problems. Among the substances most likely to pose environmental hazards at that point are lead, used for wiring in the cells; brominated flame retardants (BFRs), polybrominated biphenyls (PBBs), and polybrominated diphenyl ethers (PBDEs), used to reduce the flammability of cells; and hexavalent chromium (Cr^{6+}), added to cell coatings to improve absorption of solar radiation. All of these substances are known hazards to human health (Mulvaney 2009, 20–21).

In an effort to increase public awareness of the potential environmental risks posed by the manufacture of solar PV cells, SVTC has developed a *Solar Scorecard*. The scorecard ranks companies that manufacture solar cells on 12 categories relating to their awareness of environmental issues and their efforts to deal with those issues in their operations. The categories include topics such as emissions transparency; chemical reduction plan; workers' rights, health, and safety; module toxicity; energy use and greenhouse gas emissions; and high-value recycling. In its 2014 rankings, SVTC gave its highest ranking to the Chinese company Trina Solar, which scored 92 points out of a possible 100. Other high-ranking companies were SunPower, of Palo Alto, California (88 points), Yingli Solar, of China (81 points), SolarWorld, of Germany (73 points), and REC, of San Luis Obispo, California (71 points). The lowest-scoring companies were JA Solar, of China (10 points), Hanwha SolarOne, of China (10 points), and Canadian Solar (14 points) ("2014 Solar Scorecard" 2015).

Land Use

Every type of energy generation system requires a certain amount of land on which to operate. For some technologies, that land area can be quite small, as is the case with most coal and oil and nuclear plants. According to one frequently quoted study, these facilities often require no more than about 6 m²/MW/yr (square meters of land per megawatt-hour per year) for their operation. Natural gas plants require even less space, generally less than one square meter of land per megawatt-hour per year. By contract, renewable resource facilities, such as those used for the generation of biomass, wind, or solar power, require much larger amounts of land, nearly 500 m²/MW/yr, 100 m²/MW/yr, and 15 m²/MW/yr, respectively (National Research Council, Panel on Electricity from Renewable Resources; National Academy of Sciences; and National Academy of Engineering 2010, 213–215).

These figures provide only general comparisons, however, as they ignore a number of important relevant factors. For example, even though fossil fuel plants usually take up very little physical space, mining of the materials with which they operate—coal, oil, and natural gas—often requires the disturbance of very large areas of land. Also, the space occupied by renewable energy sources varies considerably, depending on the type of facility involved. For example, a rooftop solar or wind system has no physical footprint, so it occupies zero land space, while some commercial solar PV or CSP facilities may occupy hundreds or thousands of square meters.

It is, of course, the larger solar facilities about which some land use specialists are most concerned. Those authorities worry that such facilities will interfere with a variety of other activities, such as agricultural fields, grazing areas, military bases, mineral operations, recreational areas, and wilderness regions. Even if an area is not physically damaged, it may essentially become unavailable for any or all of these other uses. Even with a minimum of physical disturbance, some critics say that solar plants can severely interfere with the visual appearance of a natural area (Tsoutsosa, Frantzeskakib, and Gekas 2005). Indeed, some knowledgeable observers have suggested that the single most important restriction on the development of renewable energy resources such as solar energy may be the availability of land resources and the costs of developing those resources for energy production (Pimentel, et al. 2002).

Mitigation Technologies

"For every problem, there is a solution." That quotation is attributed to American satirist H. L. Mencken, who then went on to add, "which is simple, neat, and wrong." Certainly proponents of solar energy are constantly looking for solutions to the type of environmental problems described above. For example, one way of dealing with land use problems for large solar facilities is to site them in desert areas where there may be few confounding factors to disrupt the solar plant's operation. The added benefit, of course, is that such areas often have a vast supply of solar energy with which the plant can operate (Tsoutsosa, Frantzeskakib, and Gekas 2005).

Another type of location where solar facilities can be constructed without damaging land is on top of brownfields or waste disposal sites. A brownfield is a location that was previously used for industrial or commercial purposes and, as a result, has become so contaminated that it has few, if any, present uses. A solar PV array can be constructed on top of a brownfield, converting that otherwise useless area into a productive source of electricity. Similar facilities can also be built on top of landfills and other waste disposal sites that are no longer in use and are serving no other useful purpose ("Toxic Land Generates Solar Energy" 2015; "Solar Power on Landfills" 2015).

Visual effects can also be reduced by incorporating solar facilities into a variety of existing structure, such as the facades and roofs of buildings. One of the many imaginative ways to achieve this objective is the development of solar roadways, projects in which solar cells are used as the basis for roads, parking lots, and other vehicle structures ("Solar Roadways" 2015; see also Chapter 3 of this book, The SolaRoad Project.

Water Use

Water use can be an environmental problem for some types of solar technology but not others. For example, solar PV systems use essentially no water in their operation, except for processes that involve only minimal water use, such as the cleaning of mirrors and other components of the system. By contrast, CSP systems that use water for cooling (so-called *wet cooled* systems) can use up to almost 1,000 gallons of water per megawatt-hour of power produced in their operation and an additional 40–60 gallons per megawatt-hour for other purposes (again, such as cleaning the components of the system). CSP systems that are cooled by air (*dry cooled* systems) use no water in their operation but about 30–80 gallons per megawatt-hour for maintenance purposes (Hand et al. 2012, Table 10–8, 10–47). Water use issues are, unfortunately, most common in hot, dry areas, which are also characterized by a lack of natural water resources.

The most common recommendation for dealing with water issues is the conversion of solar power systems from water-cooled to air-cooled systems, where such an option is available.

Other Environmental Effects

Few or no additional environmental effects are to be expected from the operation of a solar power plant. For example, one review of such effects has concluded that "[t]here are no direct air emissions from operating a solar facility" ("Solar Energy Operations Impacts" 2015). Other potential issues of concern depend, to a considerable extent, on the design of a plant, its location, and the way it is operated. For example, some neighbors of solar facilities might complain about noise problems arising from devices such as steam turbine generators, pumps, cooling towers, solar tracking devices, and solar dish engines; transformer and switchgear noise from substations; corona noise from transmission lines; vehicular traffic noise; and maintenance facility noise ("Solar Energy Operations Impacts" 2015). Generally noise problems are likely to be much less severe than those reported for some types of wind farms.

Concerns about the disruption of archaeological, paleonto-logical, cultural, historic, and other sites of importance have also been expressed. Such concerns can be easily resolved early in the planning process, however, by ensuring that construction does not occur in locations where such resources are placed at risk. (An example of the type of environmental impact statement that might be required during the permitting process for the construction of a solar facility can be found at "Archaeological Impact Assessment for the proposed Bosjesmansberg PV East Solar Energy Facility, Located Close to Copperton in the Northern Cape" 2013.)

Net Environmental Effects

This rather lengthy section on possible environmental effects of solar energy facilities might give one the impression that such effects are a matter of great concern and a serious hindrance to the development of solar power in the United States and the rest of the world. Such an inference is probably not usually correct. While it is true that one or another facility may pose special environmental problems, many experts tend to believe that the environmental benefits of solar power far exceed the possible risks posed by that technology. For example, one of the early reviews of the costs and benefits of solar energy technologies (SETs) concluded:

> SETs present tremendous environmental benefits when compared to the conventional energy sources. In addition to not exhausting natural resources, their main advantage is, in most cases, total absence of almost any air emissions or waste products. In other words, SE can be considered as an almost absolute clean and safe energy source. (Tsoutsosa, Frantzeskakib, and Gekas 2005, 295)

Some effort has been made, in fact, to quantify the environmental benefit offered by substituting solar energy for energy produced by conventional means, such as fossil fuel or nuclear power plants. In one such study, the annual monetary value of the environmental effect alone of using solar power in preference to one of these conventional sources in the states of New Jersey and Pennsylvania was found to range from $20 to $48 for residents of New Jersey and $48 to $129 for residents of Pennsylvania (Perez, Norris, and Hoff 2012, 44).

Economic Issues

If there is any one slogan that could be used to describe the history of solar energy in the United States and throughout the world until the very recent past, it would probably be "It costs too much." Various governments at various times in history since the 1950s have taken a variety of positions about solar energy, either encouraging and supporting its development (as during the Arab oil embargo of the 1970s) or acknowledging its theoretical importance but dismissing its practical value (as during periods of low oil prices). But for all the debates and discussions over the role of solar energy in the world's energy equation, the bottom line has consistently been the fact that the cost of producing a single watt of electricity or Btu of thermal energy using solar radiation has been far greater than producing a watt by coal-, oil-, or natural gas–fired power plants or even by nuclear power plants. So the many environmental advantages of solar power have generally counted for next to nothing because of the cost of producing solar energy on a commercial scale.

Federal Subsidies

In fact, the only reason that research on solar energy was able to continue during the last quarter of the twentieth century and the first half-decade of the twenty-first century was the willingness of the federal governments in a few nations around the world (such as the United States) to provide at least some type of direct or indirect funding to such research. The 1970s were a particularly promising period in the United States for the support of solar research. The Solar Energy Research, Development, and Demonstration Act of 1974, for example, laid out an optimistic view of the potential contribution of solar energy to the nation's future energy equation. In its statement of findings in the act, the Congress reflected the view of solar energy that President Jimmy Carter had earlier expressed in his speeches on renewable energy, solar energy in particular. "It is in the Nation's interest," the Congress said, "to expedite the long-term development of renewable and nonpolluting energy resources, such as solar energy." In order to encourage solar research, the act goes on to say, "The need to make clean and renewable energy alternatives commercially viable require that the Nation undertake an intensive research, development, and demonstration program with an estimated Federal investment which may reach or exceed $1,000,000,000" ("Public Law 93–473" 1974, Section 2).

One of the shortest sections of the act was ultimately one of its most important, creating a Solar Energy Research Institute, which, according to the document, says, "shall perform such research, development, and related functions as the Chairman may determine to be necessary or appropriate in connection with the Project's activities under this Act or to be otherwise in furtherance of the purpose and objectives of this Act" ("Public Law 93–473," Section 10). The Institute (SERI; later the National Renewable Energy Laboratory) was established in Golden, Colorado, with a substantial budget and research staff for the design and conduct of research projects on solar energy.

Only a year after SERI began operations, the U.S. Congress adopted two additional acts that long served as the basis for ongoing support of solar energy research and development. The first of those acts, the Public Utility Regulatory Policies Act (PURPA), required that utility companies purchase electricity from so-called qualifying facilities, such as private homes with their own solar or wind electricity producing facilities. The act also specified that electric utilities had to pay rates for this electricity that were "just and reasonable" and that did not "discriminate against qualifying cogenerators or qualifying small power producers" ("Public Law 95–617," Section 210). The act thus became the basis for net-metering operations that have long made possible the installation and operation of small-scale solar systems. Indeed, PURPA provisions were eventually responsible for the production of about 7 percent of all electricity produced in the United States, or 71 GW of electrical energy (Maloney 2012).

The second energy bill adopted in 1978 by the U.S. Congress was the Energy Tax Act of that year. The act was sometimes known as the "gas guzzler" tax because of its attempt to have consumers cut back on the use of fuel-inefficient cars. But it had many other important ramifications. Probably the most important with regard to solar energy was the provision for tax credits to individuals and companies for investment in renewable forms of energy, such as solar and wind energy. Such tax credits are generally known as investment tax credits (ITCs) because they are given for the investment that one makes in the construction of an energy facility, not on how much power that facility actually produces. In one section of the act, a tax credit of 30 percent was given to residential use of solar energy and, in another section, a credit of 10 percent for commercial use of solar energy ("Public Law 95–618," Sections 101 and 301). Those tax provisions were scheduled to expire in either 1982 or 1985, depending on the type of credit, but the Congress decided to extend both residential and commercial tax credits, in at least one case, after the tax had already expired. The Congress then continued to renew the tax credits in one form or another annually or after every few years, often at the very last moment before they were due to expire. As a result the renewable energy industry has traditionally been left somewhat on tenterhooks as to what it could or could not expect by way of tax benefits in coming years (Gielecki, Mayes, and Prete 2001; Sherlock 2011).

The long-standing tax credits for both residential and commercial solar power facilities are still in effect and are currently set to expire on December 31, 2016. There is no way of telling as to whether those tax credits will be renewed once again, as they have so often in the past (for current information on tax credits, see "The Solar Investment Tax Credit" 2015).

No discussion of tax policy is ever very simple. In the case of solar energy, a complication that has arisen is the result of the creation of production tax credits (PTCs) in 1992. Production tax credits are based on the amount of energy produced by a facility, not on the costs of building or operating that facility. Solar energy was added to the list of technologies that could qualify for PTCs in the American Jobs Creation Act of 2004. Only a year later, however, that policy was revised with the passage of the Energy Policy Act of 2005, which removed solar energy from the list of resources that qualify for production tax credits. Although there is some interest in reversing that policy again, solar energy currently qualifies only for ITCs and not PTCs ("Renewable Energy Production Tax Credit [PTC]" 2015; Sherlock 2014.

So, the bottom line is, how effective have tax incentives been to the development of solar energy in the United States? Quantifying the effect of tax policy is a difficult challenge, but many experts in the field believe that the rather remarkable growth in solar energy projects since 2005 is in large part due to such incentives. In 2014, Ken Johnson, vice president of the Solar Energy Industries Association (admittedly not an unbiased observer), offered his view on the topic. "Annual solar installations in 2014," he said:

> will be 70 times higher than they were in 2006. . . . By the end of this year, there will be nearly 30 times more solar capacity online than in 2006. We've gone from being an $800 million industry in 2006 to a $20 billion industry today. The price to install a solar rooftop system has been cut in half, while utility systems have dropped by 70 percent. It's taken the U.S. solar industry 40 years to install the first 20 GW of solar. Now, we're going to install the

next 20 GW in the next two years. And finally—during every single week of this year—we're going to install more capacity than what we did during the entire year in 2006.

So anyway you look at it, the ITC has played a major role in solar's remarkable growth in the United States. We are more competitive; the marketplace is maturing; financing has become more innovative; and solar is no longer considered a nice, little niche industry. (Finzel 2014)

Tax Credits for Solar: A Different View

Many observers have suggested that tax incentives by the federal and state governments have been a good policy, not only for the solar industry, but for the nation as a whole. Such incentives have made possible, they say, a vigorous growth in the solar industry that will make it possible for the United States to break its dependence on fossil fuels, along with the nation's dependence on other countries for its fuel needs and its growing contribution to global climate change. But not everyone agrees with that point of view. Other observers argue that using tax dollars to support private businesses, as is the case with solar energy incentives, is an unnecessary and inappropriate intrusion of government into the marketplace. These critics say that industries should be allowed to survive or fail on their own merits and not be allowed to rely on the government for their ultimate fate.

During the second decade of the twenty-first century, that position became particularly popular among some elements of the Republican Party, which in any case tends to stress the importance of the marketplace in determining the direction of the nation's economy. One of the specific events to which Republicans drew particular attention was the failure of the Solyndra corporation, a manufacturer of CIGS solar cells. Solyndra went bankrupt in 2011 after defaulting on a $536 million loan from the U.S. Department of Energy (DOE). A number of Republican critics suggested that Solyndra's failure was only the most recent example of the federal government's interference in the marketplace, an attempt to prop up a company with tax incentives that could not otherwise survive on its own. For example, Republican presidential candidate Mitt Romney mentioned on a number of occasions that Solyndra was a glaring example of President Barack Obama's misguided energy policies. In one of his ads, Romney said that solar energy tax policy was "just another example of President Obama's pattern of picking winners and losers and wasting taxpayer money" (Geman 2012). Supporters of solar energy tax policy have responded to such attacks by pointing out that overall, such loan guarantee programs have actually made a significant profit for the federal government (see, for example, Groom 2014).

The increasing influence of the Republican Party at both federal and state levels as a result of the 2014 elections began to have some influence on tax policy with regard to solar and other forms of renewable energy. For example, the state of Florida changed course on renewable energy in 2014, when it decided to reduce the state's energy efficiency goals by 90 percent and to end the state's solar rebate program in 2015. The decision by the state Public Utility Commission (PUC) appeared to reflect the views of Governor Rick Scott, who had previously opposed a bill offering tax credits for solar energy because they provided no concrete benefits to citizens of the states (Taylor 2013).

There is some reason to believe that the actions taken by the Romney campaign in 2012 and the Florida PUC are indicative of the way that most Republicans in general feel about solar energy tax policy. In its most recent poll on the question, the Pew Research Center for the People & the Press reported that support for public funding of renewable energy projects had fallen from 83 percent in 2006 to 53 percent in 2011 among respondents who identified themselves as Republicans or leaning-Republican. During the same period, support for this type of funding remained essentially level (81 to 83 percent) among Democrats.

Political observers are currently asking how this shift in support for solar energy tax policy among Republicans in general is likely to affect national policy in coming years. One view is that Republican legislators, who now control both houses of Congress, are more likely to reverse the course of federal funding for renewable energy in coming years (Cardwell 2015). Other observers are, however, optimistic that the Congress will continue to provide at least some type of support for new projects in solar, wind, geothermal, tidal, and other forms of renewable energy. Perhaps the best bet will be that legislators at both federal and state levels, no matter their political predispositions, are likely to continue battling out the best way to promote solar energy over the next decade.

State Programs

The vast majority of incentives for the development and use of solar energy in the late 1970s came from federal programs like those described earlier. However, similar efforts were also under way in a handful of states and local communities. Most prominent among these efforts were a series of legislative and administrative actions in the state of California designed to promote a significant increase in the use of renewable energy sources, solar among them. In 1978, for example, the state legislature considered a bill with the objective of establishing a plan for "'maximum feasible solarization' of California in space, and water heating and photovoltaics, by 1990" (Elliott 1978, 17). The legislature also considered some of the earliest sunlight-protection laws designed to ensure "legal protection for the right to receive sunlight necessary to operate solar energy systems" (Elliott 1978, 17).

California governor Jerry Brown was also moving on his own initiative to devise ways of promoting solar power. In 1978 he created a state commission called SolarCal, reportedly based on suggestions from political activist Tom Hayden, designed to encourage the development of solar energy facilities in the state. One observer called the new organization a kind of "alternative energy commission" (Elliott 1978, 17). One of the tools with which the agency had to work was a very generous tax credit of 55 percent for the purchase of solar energy equipment, which was generally credited with providing a significant boost to the solar energy industry in the state.

Although the most ambitious solar energy programs in the nation, California's efforts were not the first such programs. For example, as early as 1976, the state of Hawaii enacted a 35 percent income tax credit for the cost of equipment and installation on both single-family and multifamily residences in the state. Those credits have remained in effect, with some changes in tax credit amounts, to the present day ("Solar and Wind Energy Credit" 2014; "House Bill §235-12.5" 1976).

Over time, various states implemented a variety of plans for rewarding the purchase and use of solar facilities, including credits on one's personal income tax, property tax exemptions, business tax deductions, sales tax exemptions, accelerated amortization, and, in one case (Wisconsin), cash payment for the adoption of solar energy use (Roessner 1980; Sarzynski 2009). Overall, the first tax incentives for solar energy were adopted by Arizona and Indiana in 1974. In both states, those incentives were property tax credits. By 1976, 26 more states had adopted some form of financial incentive for the construction and/or use of solar energy, and by 1981, that number had reached 44 states for some type of personal tax credit (Sarzynski 2009, 18). Today, the most popular type of financial credit provided by states for the use of solar facilities is a property tax credit, followed by some form of financial aid for construction of facilities, cash payments (including grants, loans, and rebates), corporate tax credit, personal income tax credit, and sales tax credit (Sarzynski 2009, Figure 1, 9).

The Role of Big Business

The early history of solar energy is commonly a story of an individual or a small group of researchers, often working in their own private laboratory with their own personal financing, developing improvements and modifications in basic equipment such as solar cells and solar collectors. By the late

1970s, however, that model had begun to change. In part because of the federal government's increasing interest in solar energy, along with its willingness to spend millions of dollars on product design and development, larger energy companies had become involved in the solar energy development market. As a result, the decade saw most—and eventually all—of the small start-up solar energy companies being swallowed up by energy giants such as Mobil, Exxon, Atlantic-Richfield, and Shell Oil. The motivation behind these acquisitions was not always entirely clear. In some cases, a company clearly envisioned practical applications for solar cells in their operations, as in their use as power sources on offshore platforms. In other instances, the giant energy companies may also have begun to appreciate the possibility of long-term changes in the way the world would produce and use energy, with solar and other forms of renewable energy becoming an increasingly factor in that change.

In any case, the 1970s saw the gradual absorption of small start-up companies by the largest energy and manufacturing companies in the world. In a review of this process, Harvard Business School researchers Geoffrey Jones and Loubna Bouamane summarized some of the most important changes during the period (Jones and Bouamane 2012). For example, researcher Elliot Berman had developed a more efficient type of solar cell in the 1970s but was unable to obtain financing on his own to produce the cells. In 1973, he accepted an offer from the oil giant Exxon to create his own company, the Solar Power Corporation, which was, however, a wholly owned subsidiary of Exxon. Four years later, Exxon sold Solar Power to ARCO, which, in turn, sold the company to an Amoco-owned company called Solarex in 1979, after which Solar Power was sold once more in 1990 to Siemens. This type of "musical chairs" transfer of smaller companies among major energy and manufacturing corporations was increasingly common during the 1970s, when small solar companies were struggling to survive and larger entities were struggling to figure out how to include solar energy in their own portfolios of activities (Jones and Bouamane 2012, 21–29; "Big Business Squeezing Out Small Solar Firms—Lobbyists" 1980, 8C).

The Pros and Cons of Solar Energy

One point to be learned from a review of the history of solar energy is that some disagreement remains as to the potential benefits and risks, advantages and disadvantages, of this form of technology. Some observers see almost unlimited potential for solar energy in solving a host of the world's problems, while others see some serious problems with the increased use of the technology. Here are some of the arguments in support of and in opposition to solar energy.

Pros

- Solar energy is a form of renewable energy. Unlike the case of fossil fuels such as coal, oil, and natural gas, there is no possibility that humans will run out of solar energy for the foreseeable future (at least a few billion years).
- Solar energy is abundant. More solar energy falls on Earth in one day than all of humanity uses in one year ("Solar Energy Basics" 2014).
- Solar energy is available over most parts of Earth. In contrast to deposits of fossil fuels, which tend to be concentrated in a relatively few specific locations, solar power can be produced just about anywhere in the world with the exception of the polar regions.
- As a raw material, solar energy is free. Like other energy sources, there is a cost to collecting solar energy, but the raw material itself—solar radiation—arrives on Earth's surface at no expense to humans.
- Solar energy is reliable. The sun rises every day of the year in every part of the globe, although clouds may obscure some areas for a greater or lesser part of the day.

- Solar energy is environmentally friendly. Unlike other sources of energy, such as fossil fuels and nuclear power, solar energy releases very few pollutants to the air, water, and land. Perhaps most important in today's world, solar energy releases no greenhouse gases to the atmosphere and, therefore, contributes virtually nothing to global climate change.

- Solar energy is sustainable; that is, it can be used to meet the energy needs of today's world without placing at risk its possible use in the future.

- Solar energy can reduce electricity costs. The process of net metering means that homes, offices, and other structures that produce their own electricity can sell that excess energy to utilities, thus reducing their own utility bills.

- Although construction costs for solar energy facilities are still high, operating expenses for solar PV facilities are at least equal to or less than such costs for almost all other types of electricity-generating facilities. About all that's needed to keep a solar PV facility operating is a wash-down of cells about once a month ("Updated Capital Cost Estimates for Utility Scale Electricity Generating Plants" 2013, Table 1, 6).

- The cost of energy derived from solar radiation has been decreasing by substantial amounts over the last decade, and indications are that that trend is likely to continue into at least the near future ("Solar Voltaic Roadmap" 2010).

- Various types of financial incentives are available for the purchase, installation, and use of solar facilities. The federal government, many states, and some local municipalities offer tax credits, loans, outright grants, sales tax deductions, and other financial benefits for using solar energy, benefits that are not available for some other types of energy production.

Cons

- The estimated levelized cost of electricity (LCOE) for new electricity plants entering service in 2019 is estimated to be either 130.0 for new solar PV plants or 243.1 for new solar thermal plants, much higher than the value for conventional coal plants (95.6), natural gas plants (as low as 64), nuclear plants (96), and onshore wind farms (80). LCOE is a measure of all costs associated with building, operating, and dismantling a plant. These numbers show that solar power is still very expensive compared to other types of power generation ("Levelized Cost and Levelized Avoided Cost of New Generation Resources in the Annual Energy Outlook" 2014).

- Solar radiation is intermittent. It varies as the seasons change and as day becomes night, and as weather conditions permit greater or lesser amounts of sunlight to reach the ground. This very basic fact means that some method or methods must be found for storing solar energy when the sun is not shining. Although progress is being made, storage systems are still very expensive to build and maintain (for a discussion of this point, see Vorrath 2015).

- The use of solar energy may be associated with a number of environmental issues, as discussed in some detail in preceding sections (see also Gunerhan, Hepbasli, and Giresunlu 2009; Hernandez, et al. 2014; Kaygusuz 2009; Tsoutsosa, Frantzeskakib, and Gekas 2005).

- Disadvantages of solar power differ to some extent from technology to technology. For example, CSP systems tend to be quite complex and, therefore, relatively expensive, especially when compared to solar PV technologies. This fact explains the very high LCOE for solar thermal plants compared, for example, to solar PV and conventional coal-fired plants (Crawford 2013). By contrast, the "working part" of solar PV systems, the solar cells of which they are composed, tend to be fairly simple. But the connections needed to carry electricity generated by the cells to the ultimate source may require a somewhat complex system of wires, inverters, and storage devices.

Public Opinion about Solar Energy

The growing popularity of solar energy among consumers and producers mentioned earlier in this chapter is also reflected in public opinion surveys on the topic. Definitive data from these surveys is not always available for extended periods of time because public opinion surveys until very recently tended to lump all forms of alternative or renewable energy together in the questions they asked respondents. Nonetheless, it appears quite clear that the vast majority of Americans have long accepted the fact that all forms of renewable energy, including solar, should be a significant part of the nation's energy equation and, usually, should receive greater emphasis from the government than being given at the time of the survey in question.

A review of the literature conducted for the National Renewable Energy Laboratory in 2011, for example, found that those responding to surveys over the preceding decade always responded positively in a majority of at least 80 percent to questions about "the use of renewable energy sources." That majority declined somewhat over the decade, dropping from 89 percent who "completely" or "somewhat" agreed with the proposition in 2002 to 80 percent in 2010, but always consisted of a majority of between 41 and 57 percent for those who supported renewable energy "completely" ("Consumer Attitudes about Renewable Energy: Trends and Regional Differences" 2011).

In recent years, surveys tend to focus more specifically on public attitudes about solar energy in conjunction with other forms of renewable energy. For example, a 2013 survey by the Gallup organization found that 76 percent of all Americans agreed that the United States should place "more emphasis" on solar energy; with wind energy receiving the same response from 71 percent of respondents, natural gas, 65 percent; oil, 46 percent; nuclear power, 37 percent; and coal, 31 percent. Favorable levels of response for solar energy were recorded for both political parties (Democrats, 87 percent; Republicans, 68 percent) and all regions of the country (East, 79 percent;

Midwest, 75 percent; South, 74 percent; West, 78 percent) (Jacobe 2013). These data are roughly the same for almost every other country in the world where similar questions have been asked about the role of solar and other forms of renewable energy in national economies (see, especially, "Public Opinion on Global Issues" 2012, 14–15).

Solar Power Today

"Solar power has been declared a winner before, only to flounder." An early 2015 article in the prestigious journal *Foreign Affairs* began with this warning (Pinner and Rogers 2015). The article then went on to say, however, that the development of solar energy in most parts of the world has been remarkably impressive in the past decade. Progress in the construction of new solar facilities has blossomed not only in developed nations, such as Germany, Spain, the United States, and Japan, but also in smaller, less developed nations in nearly every part of the world. The government of India, for example, is hoping that solar energy will be the answer to providing electrical energy to more than 100,000 small villages that currently do not have access to electricity. A number of challenges remain, the article continued, in order for solar energy to achieve its potential, including better regulatory policies and unexpectedly robust competition from natural gas, nuclear energy, or some other energy source. But, the authors concluded, most signs seem to be positive for a continued growth of solar energy, and it appears that "the momentum behind solar power has become unstoppable" (Pinner and Rogers 2015).

At the front of growth in solar energy is a somewhat surprising leader, China. As the world's most populous nation, producing the greatest amount of air pollution of any country, China would hardly seem to be a candidate for leader of the world's efforts to develop renewable energy sources. But such is the case, and now has been for the past half-decade or so. In fact, the Chinese are now the world's largest producer of solar PV panels, turning out two-thirds of all panels built in the world in 2014. Although still lagging slightly behind Germany in terms of installed solar facilities, 23 gigawatts of energy compared to 36 gigawatts of energy, China now spends more on solar research—$23.5 billion in 2014—than any other nation in the world. All of these developments, along with the promise of even greater growth, are the result of the Chinese government's decision to make a long-term commitment to the switch from fossil fuels to renewable energy (Hersh 2015).

At the same time, progress in the development of solar energy in the United States is somewhat less certain. A critical moment arrived in 2015 with the question as to whether or not federal subsidies for solar energy would be renewed at the end of 2016. Most observers believe that the remarkable progress in solar development in the United States over the past decade owes a great deal to the generous benefits provided by federal and state governments to solar power users. And the question becomes to what extent that development can continue if subsidies are withdrawn or reduced. Although the role of solar energy in the nation's energy equation in the long term seems assured, there is less certainty as to how the American industry could adapt to a withdrawal of the major financial impetus provided by government incentives.

One indication of this uncertainty is the (temporary?) halt in the construction of the very largest solar facilities, called *solar farms*, in the United States. Between 2010 and 2015, 17 of these huge facilities were constructed in the United States, with nominal power capacities ranging from about 60 megawatts to 579 megawatts (Solar Star, in the western Mojave Desert in California) ("The World's 20 Largest Solar Projects" 2015). But no solar project of similar size is currently planned, primarily, according to some observers, because of the uncertainty as to how or to what extent the federal government will be able to offer financial incentives for the cost of such a facility and whether or not the cost of solar electricity will continue to decrease (Hobbs 2015, but see responses to the author's analysis of the current situation at the conclusion of this article).

The primary research program for the development of solar energy in the United States today is SunShot, sponsored by the DOE. SunShot was conceived in 2010 as a program for sponsoring research, development, education, and related projects that could reduce the cost of solar electricity by 75 percent compared to 2010 prices by the year 2020. The program is an ambitious effort that involves research grants, financial awards, fellowships, publications, and other methods for advancing interest in and research on solar photovoltaics, concentrating on solar power, systems integration, and related fields of solar technology. The DOE funded 14 SunShot projects in its first year of operation (2011), nine more in 2012, eight in 2013, and six in 2014. Examples of the types of research supported by SunShot include rooftop solar facilities, integration into the electric grid, market factors in the use of solar power, improving the accuracy of solar forecasting, small business research in solar energy, and next-generation solar technology ("SunShot Initiative" 2014; "SunShot Initiative" 2015).

The Future of Solar Power

So what is the outlook for solar power over the next few decades? Opinions differ, of course, about the answer to that question, depending on whom one asks. Certainly, companies involved in the production of solar power technology tend to be very optimistic as to where solar power will stand 10, 20, 30, or more years from now. But it appears that many experts and associations interested in energy issues tend to share those views in general. For example, the highly respected Bloomberg New Energy Finance (BNEF) report for 2014 predicted a rosy future for solar technology into at least 2020 and 2030. The report predicted that rooftop solar photovoltaic systems would make up about a fifth of all the new energy installations created by the year 2020. Overall, BNEF 2014 said, solar energy would account for by far the greatest fraction of new energy installations between 2015 and 2030, exceeding both fossil fuel and other renewable energy sources. Whereas solar facilities constituted only about 2 percent of the world's installed energy capacity in 2012, BNEF predicted that the number would rise to 18 percent by 2030. By comparison, the report suggested that wind capacity would increase only from 5 percent in 2012 to 12 percent in 2030 and other renewables would actually decrease from about 22 percent in 2012 to 19 percent in 2030 (fossil fuels were predicted to drop from 64 percent of the world's installed capacity in 2012 to 44 percent in 2030; "2030 Market Outlook" 2014, Global Overview).

There now appear to be two major trends with regard to the direction of solar energy in the near future. First, countries that have already made a commitment to the role of solar energy appear to have concluded that decision was a good one, and they expect to expand their investment in energy from the Sun. The BNEF report said that by 2030, for example, Germany would be getting just over half of all its energy (52 percent) from those two renewable energy sources alone, wind and solar ("2030 Market Outlook 2014," Capacity by Technology). This trend is perhaps most obvious in Asia, where at least three of the region's largest countries—Japan, China, and India—have made a commitment to increasing the role of solar energy in their energy equations. In Japan, which has essentially no fossil fuel reserves of its own, interest has turned to renewable sources of energy, especially solar, since the Fukushima nuclear disaster of 2011 soured many Japanese views on the nation's dependence on nuclear power. In the four years following the Fukushima event, solar facilities began springing up in every conceivable spare space on the island nation, including rice fields, golf courses, unused highways and airports, and vast stretches of near-land ocean waters (Soble 2015). And in 2015, Japanese researchers made an important breakthrough in the technology needed to transmit solar energy from space satellites to Earth-based receiving stations ("Japan Space Scientists Make Wireless Energy Breakthrough" 2015). The relative importance of nuclear and solar energy is still a matter of some dispute in Japan, but the likelihood of continued growth in the solar field now seems assured.

Enthusiasm for solar energy seems even more pronounced in China. Although that nation has long been the world's leading producer of solar photovoltaic cells, the vast majority of that product has always gone to exports to other nations. Over the past decade, however, the Chinese government has taken a second look at its own solar program and decided that it needs to make a greater commitment to its own domestic use of solar energy. Indeed, the story of solar energy in China has been one of constantly increasing solar energy goals since the adoption of the nation's Tenth Five-Year Plan in 2000. During succeeding years, the national government repeatedly announced that stated goals for solar energy were too pessimistic and that the country was actually producing significantly more solar energy than had been predicted. In 2009, for example, Wang Zhongying, head of China's National Renewable Energy Development Centre, announced that the country was likely to be producing at least five times as much solar energy (about 10,000 megawatts) as the goal set for the nation only two years earlier (1,800 megawatts).

That trend has continued to the present day. In 2013 (the most recent year for which data were available), China installed just more than 12,000 megawatts of solar PV facilities, almost twice as much as the next largest producer of solar energy, Japan, and more than three times as much as the United States. The new figure for installed capacity even reflected an increase of 232 percent over China's own figure for the preceding year ("Global Renewal Energy Report 2014" 2014). A number of reasons explain China's continuing and growing commitment to the use of solar (and other renewable) energy, one of the most important of which was the nation's increasingly severe problem of air pollution.

India faces many of the same energy problems as do Japan and China, with a huge population, a quarter of whom have no access to utility-scale electricity, living in a country with massive air pollution problems. In late 2014, Prime Minister Narendra Modi announced a new initiative to expand its solar energy capacity by 30 times by the year 2020. Thus far, however, the nation has made relatively little progress in even approaching such a goal, and critics worry that internal Indian politics will make it difficult for the nation to achieve Modi's ambitious target (Fairley 2015).

The second trend in the future of solar power worldwide appears to be the growing willingness of nations that have largely ignored the development of renewable resources to begin exploring the harvesting of solar energy. Perhaps the most noteworthy example of this trend is Saudi Arabia, a nation with the world's largest proved reserves of fossil fuels, and not one, therefore, to be expected to have much interest in the development of renewable resources. Such has turned out not to be the case, however. At the fourth Solar Arabia summit held in Riyadh, Saudi Arabia, in October 2014, for example, Hamed al-Saggaf, executive director of the Saudi Electricity Company, told delegates that his nation was making a very serious commitment to the development of solar power over the next few decades. It had already made plans to spend $109 billion for the development of solar facilities, with plans for solar energy to produce 30 percent of the nation's electricity by 2032. Saudi Arabia was aware, al-Saggaf said, that its fossil fuel resources would not last forever and that it would be wise for the nation to start planning for the post-fossil-fuel era as soon as possible. He noted that the nation's rulers had come to the conclusion that solar had now become "a must" for the nation's future (Clover 2014).

Another very different example of this trend is the small African nation of Rwanda. In May 2015, the Rwandan government opened East Africa's first utility-scale solar power facility. The new plant is designed to produce 8.5 megawatts of electricity, enough to power 15,000 homes, meeting about 5 percent of the nation's energy needs. At the opening ceremonies for the plant, Minister of Infrastructure James Musoni noted that his country was committed to planning for its future energy needs with a mix of hydroelectric, peat, methane, solar power, and other resources. He said the new solar facility would be an important element in providing "clean and reliable energy to satisfy growing demand and sustain national development" ("East Africa's First Utility-scale Solar Power Plant to Be Inaugurated in Rwanda, Bringing Clean Energy to 15,000 Homes" 2015).

As important as these trends are in and of themselves, they also highlight another basic fact about solar energy: in general, it is largely unused or underutilized in the world today. Of the world's 196 nations, only about 50 have any form of solar facilities that produce significantly measurable amounts of solar power ("International Energy Statistics" 2015). Of the remaining 146 or so nations without significant solar power facilities, some seem poised to make the commitment to solar energy in the near future, while others appear satisfied to ignore that resource, at least for the foreseeable future. An example of the former case is one of the largest and fastest growing nations in the world, Brazil. At the end of 2014, the nation (with one of the world's highest rates of insolation) had an installed solar PV capacity of 15.2 gigawatts, capable of supplying about 0.1 percent of the nation's electricity needs. In early 2015, two new solar PV plants were installed with a nameplate capacity of 11 megawatts, which, according to some observers, was, at the least, a move "starting the nation down the path of utility-scale solar development" (Kessler 2015). That path was, as of 2015, however, likely to be a long and uphill struggle, not unlike that of most other nations in the world.

As discussed earlier, predicting the future of solar power in the United States is a difficult challenge. In many regards, indications are that the nation is now on a fairly rapid upswing in its commitment to all forms of solar power. The cost of solar electricity and solar heat is dropping, more facilities are being built, and many experts in the field are optimistic about the future of the technology. But challenges remain. Will the federal and state governments continue to provide financial support for a technology that may, without that support, not be able to survive on its own? Or has solar energy truly reached the turning point in its economic history and now become capable of continuing to grow and development, with or without outside support? As the old adage says, only time will tell.

References

Alsema, E. A. 1996. "Environmental Aspects of Solar Cell Modules: Summary Report." Netherlands Agency for Energy and the Environment. http://solar-club.web .cern.ch/solar-club/Textes/PVtoxic.pdf. Accessed on March 9, 2015.

"Ammonia-Based Energy Storage," 2014. Australian National University. Solar Thermal Group. http://stg.anu.edu.au/ research/storage/ammonia.php. Accessed on July 24, 2015.

"Archaeological Impact Assessment for the Proposed Bosjesmansberg PV East Solar Energy Facility, Located Close to Copperton in the Northern Cape." 2013. Heritage Contracts and Archaeological Consulting. http:// www.sahra.org.za/sahris/sites/default/files/heritagereports/ Appendix%20G%20-%20Heritage_1.pdf. Accessed on July 24, 2015.

"Arsine (SA): Systemic Agent." 2014. Centers for Disease Control and Prevention. http://www.cdc.gov/niosh/ershdb/ emergencyresponsecard_29750014.html. Accessed on March 6, 2015.

"AzRISE Energy Storage Initiative." 2012. Arizona Research Institute for Solar Energy. http://www.azrise.org/ azrise-research-overview/azrise-energy-storage-initiative/. Accessed on March 4, 2015.

"Big Business Squeezing Out Small Solar Firms—Lobbyists." 1980. Lakeland Ledger. http://news.google.com/newspa pers?nid=1346&dat=19801115&id=g0dNAAAAIBAJ& sjid=IPsDAAAAIBAJ&pg=6854,5164515. Accessed on March 11, 2015.

Braun, Andrew. 2014. "Solar Power: Energy of the Future?" IDG Connect. http://www.idgconnect.com/

abstract/9013/solar-power-energy-source-future. Accessed on February 28, 2015.

Cardwell, Diane. 2015. "Worry for Solar Power after End of Tax Credits." *New York Times.* http://www.nytimes.com/2015/01/26/business/ worry-for-solar-projects-after-end-of-tax-credits.html?_r=0. Accessed on March 13, 2015.

"Challenges and Opportunities for New Pumped Storage Development." 2012. Pumped Storage Development Council. http://www.hydro.org/wp-content/ uploads/2012/07/NHA_PumpedStorage_071212b1.pdf. Accessed on March 4, 2015.

Clover, Ian. 2014. "Solar Power Key for Saudi Future, Says Energy Chief." *PV Magazine.* http:// www.pv-magazine.com/news/details/beitrag/ solar-power-key-for-saudi-future—says-energy-chief_ 100016969/#axzz3UgK5ICwq. Accessed on March 17, 2015.

"Concentrating Solar Power Projects by Country."

"Consumer Attitudes about Renewable Energy: Trends and Regional Differences." 2011. National Renewable Energy Laboratory. http://apps3.eere .energy.gov/greenpower/pdfs/50988.pdf. Accessed on March 18, 2015.

Crawford, Mark. 2013. "Catching the Sun." *Mechanical Engineering.* 135(3): 32–37. https://www.asme.org/ge tmedia/44edaee0-d607-4ec4-b241-1b7877bdbd01/ Catching-the-Sun.aspx. Accessed on March 12, 2015.

DNV KEMA. 2013. "Residential Solar Energy Storage Analysis." http://ny-best.vm-host.net/resource/ dnv-kema-residential-solar-energy-storage-analysis. Accessed on March 3, 2015.

"East Africa's First Utility-Scale Solar Power Plant to Be Inaugurated in Rwanda, Bringing Clean Energy to 15,000 Homes" 2015. Republic of Rwanda. http://www.gov.rw/

news_detail/?tx_ttnews%5Btt_news%5D=1056&cHas
h=606e6c44d3651b267477ee1b3aa94e2e. Accessed on
March 17, 2015.

Elliott, Dave. 1978. "Solarcal." Undercurrents. http://issuu
.com/undercurrents1972/docs/__uc30_feb27a/19.
Accessed on March 12, 2015.

Fairley, Peter. 2015. "India's Ambitious Bid to Become a Solar
Power." http://www.technologyreview.com/news/535551/
indias-ambitious-bid-to-become-a-solar-power/. Accessed
on March 18, 2015.

Finzel, Ben. 2014. "Is This the End of the Line for
Wind and Solar Energy Tax Credits? Should It Be?"
OurEnergyPolicy.org. http://www.ourenergypolicy.org/
is-this-the-end-of-the-line-for-wind-and-solar-
energy-tax-credits-should-it-be/. Accessed on
March 11, 2015.

"Freeing the Grid 2015: Best Practices in State Net Metering
Policies and Interconnection Procedures." 2015. Interstate
Renewable Energy Council and Vote Solar. http://
freeingthegrid.org/. Accessed on July 24, 2015.

Fthenakis, Vasilis M., and Biays Bowerman. 2015.
"Environmental Health and Safety (EHS) Issues in
III-V Solar Cell Manufacturing." Brookhaven National
Laboratory. https://www.bnl.gov/pv/files/pdf/art_168.pdf.
Accessed on March 9, 2015.

Geman, Ben. 2012. "Romney Ad Hits Obama for 'Wasting'
Taxpayer Money on Solyndra." The Hill. http://thehill.com/
policy/energy-environment/229763-romney-campaign-
slams-obama-on-solyndra. Accessed on March 13, 2015.

Gielecki, Mark, Fred Mayes, and Lawrence Prete.
2001. "Incentives, Mandates, and Government
Programs for Promoting Renewable Energy." Energy
Information Administration. http://lobby.la.psu.
edu/_107th/128_PURPA/Agency_Activities/EIA/

Incentive_Mandates_and_Government.htm. Accessed on March 11, 2015.

"Global Cumulative Solar PV Capacity at the End of 2013, by Country (in Gigawatts)." 2015. Statista. http://www.statista.com/statistics/264629/ existing-solar-pv-capacity-worldwide/. Accessed on March 2, 2015.

"Global Installed Power Generation Capacity by Renewable Source." 2011. Today in Energy. http://www.eia .gov/todayinenergy/detail.cfm?id=3270. Accessed on February 28, 2015.

"Global Renewable Energy Report 2014." 2014. Hanergy Holding Group. http://www.hanergy.com/en/upload/ contents/2014/07/53bfa4772d9f6.pdf. Accessed on March 18, 2015.

Groom, Nicola. 2014. "Exclusive: Controversial U.S. Energy Loan Program Has Wiped Out Losses." Reuters. http://www.reuters.com/article/2014/11/13/ us-doe-loans-idUSKCN0IX0A120141113. Accessed on March 13, 2015.

Gunerhan, H., A. Hepbasli, and U. Giresunlu. 2009. "Environmental Impacts from the Solar Energy Systems." *Energy Sources Part A: Recovery Utilization and Environmental Effects.* 31(2): 131–138

Hand, M. M., et al., eds. 2012. "Renewable Electricity Futures Study," 4 vols. National Renewable Energy Laboratory. http://www.nrel.gov/analysis/re_futures/. Accessed on March 10, 2015.

Hernandez, R. R., et al. 2014. "Environmental Impacts of Utility-Scale Solar Energy." *Renewable and Sustainable Energy Reviews.* 29(1): 766–779.

Hersh, Sam. 2015. "China, the Unlikely Leader of the Green Revolution." McGill International Review. http:// mironline.ca/?p=4231. Accessed on March 13, 2015.

Hobbs, Alicia. 2015. "Are Solar Farms Declining in the United States?" Social Media Today Community. http://theenergycollective.com/aliciahobbs/2191276/ are-solar-farms-declining-united-states. Accessed on March 13, 2015.

Hoppmann, Joern, et al. 2014. "The Economic Viability of Battery Storage for Residential Solar Photovoltaic Systems—A Review and Simulation Model." *Renewable and Sustainable Energy Reviews*. 39: 1101–1118.

"House Bill §235-12.5." 1976. http://www.capitol. hawaii.gov/hrscurrent/Vol04_Ch0201-0257/ HRS0235/HRS_0235-0012_0005.HTM. Accessed on March 12, 2015.

"International Energy Statistics." 2015. U.S. Energy Information Administration. http://www.eia.gov/cfapps/ ipdbproject/iedindex3.cfm?tid=6&pid=116&aid=12&cid= regions&syid=1980&eyid=2012&unit=BKWH. Accessed on March 2, 2015.

Jacobe, Dennis. 2013. "Americans Want More Emphasis on Solar, Wind, Natural Gas." Gallup. http://www.gallup.com/ poll/161519/americans-emphasis-solar-wind-natural-gas. aspx. Accessed on March 18, 2015.

"Japan Space Scientists Make Wireless Energy Breakthrough." 2015. Phys.org. http://phys.org/ news/2015-03-japan-space-scientists-wireless-energy.html. Accessed on March 18, 2015.

Jones, Geoffrey, and Loubna Bouamane. 2012. " 'Power from Sunshine': A Business History of Solar Energy." Harvard Business School. http://www.hbs.edu/ faculty/Publication%20Files/12-105.pdf. Accessed on March 12, 2015.

Kaygusuz, K. 2009. "Environmental Impacts of the Solar Energy Systems." *Energy Sources Part A: Recovery Utilization and Environmental Effects*. 31(15): 1376-1386.

Kessler, Richard A. 2015. "EGP Starts Work on Brazil's First Utility-Scale Solar Projects." http://www.rechargenews.com/solar/1392273/ egp-starts-work-on-brazils-first-utility-scale-solar-projects. Accessed on March 18, 2015.

Kuravi, Sarada, et al. 2013. "Thermal Energy Storage Technologies and Systems for Concentrating Solar Power Plants." *Progress in Energy and Combustion Science*. 39: 285–319. http://research.fit.edu/nhc/documents/PECS. pdf. Accessed on March 4, 2015.

"Levelized Cost and Levelized Avoided Cost of New Generation Resources in the Annual Energy Outlook 2014." 2014. U.S. Energy Information Administration. http://www.eia.gov/forecasts/aeo/pdf/ electricity_generation.pdf. Accessed on March 12, 2015.

Lloyd, Robin. 2008. "Solar Power to Rule in 20 Years, Futurists Say." LiveScience. http://www.livescience .com/4824-solar-power-rule-20-years-futurists.html. Accessed on February 28, 2015.

Maloney, Peter. 2012. "PURPA: Still Generating Electricity, but with a Few Worms in the Mix." The Barrel. http:// blogs.platts.com/2012/10/09/purpa_today/. Accessed on March 11, 2015.

"Market Analysis: Current US Energy Storage Market." 2014. http://www.slideshare.net/EdgeBusinessAdvisory/ e-energy-storage-market-analysis-2014-0121. Accessed on March 4, 2015.

Matteson, Stephanie. 2015. "Batteries Shmatteries: Let's Talk about the Biggest Type of Solar Storage." The Energy Collective. http:// theenergycollective.com/stefaniematteson/2191826/ batteries-shmatteries-let-s-talk-about-biggest-type- solar-storage. Accessed on March 4, 2015.

Mendelsohn, Michael. 2012. "Where Did All the Solar Go? Calculating Total U.S. Solar Energy Production." National Renewable Energy Laboratory. https://financere.nrel.gov/finance/content/calculating-total-us-solar-energy-production-behind-the-meter-utility-scale. Accessed on February 28, 2015.

Mulvaney, Dustin. 2013. "Hazardous Materials Used in Silicon PV Cell Production: A Primer." *Solar Industry Magazine*. 6(8): 17–19. http://www.solarindustrymag.com/issues/SI1309/FEAT_05_Hazardous_Materials_Used_In_Silicon_PV_Cell_Production_A_Primer.html. Accessed on March 6, 2015.

Mulvaney, Dustin. 2014. "Solar Energy Isn't Always as Green as You Think." *IEEE Spectrum*. http://spectrum.ieee.org/green-tech/solar/solar-energy-isnt-always-as-green-as-you-think. Accessed on March 9, 2015.

Mulvaney, Dustin, et al. 2009. "Toward a Just and Sustainable Solar Energy Industry." Silicon Valley Toxics Coalition. http://svtc.org/wp-content/uploads/Silicon_Valley_Toxics_Coalition_-_Toward_a_Just_and_Sust.pdf. Accessed on March 6, 2015.

National Research Council, Panel on Electricity from Renewable Resources; National Academy of Sciences; and National Academy of Engineering. 2010. *Electricity from Renewable Resources: Status, Prospects and Impediments*. Washington, DC: National Academies Press.

Ngai, Eugene Y. 2008. "Silane Safety/Lessons Learned and Accident Prevention." Air Products and Brookhaven National Laboratory. https://www.bnl.gov/pv/files/pdf/IEEE_May2008_Silane_Tutorial.pdf. Accessed on March 6, 2015.

Perez, Richard, Benjamin L. Norris, and Thomas E. Hoff. 2012. "The Value of Distributed Solar Electric Generation to New Jersey and Pennsylvania." Clean Power Research. http://mseia.net/site/wp-content/uploads/2012/05/MSEIA-Final-Benefits-of-Solar-Report-2012-11-01.pdf. Accessed on March 10, 2015.

"Phosphine: Lung Damage Agent." 2014. Centers for Disease Control and Prevention. http://www.cdc.gov/niosh/ershdb/emergencyresponsecard_29750035.html. Accessed on March 6, 2015.

Pimentel, David, et al. 2002. "Renewable Energy: Current and Potential Issues." *BioScience* 52(12): 1111–1120

Pinner, Dickon, and Matt Rogers. 2015. "Solar Power Comes of Age." *Foreign Affairs.* http://www.foreignaffairs.com/articles/143066/dickon-pinner-and-matt-rogers/solar-power-comes-of-age. Accessed on March 13, 2015.

"Potential Health and Environmental Impacts Associated with the Manufacture and Use of Photovoltaic Cells." 2004. California Energy Commission. http://www.energy.ca.gov/reports/500-04-053.PDF. Accessed on March 6, 2015.

Poullikkas, Andreas. 2013. "A Comparative Overview of Large-Scale Battery Systems for Electricity Storage." *Renewable and Sustainable Energy Reviews.* 27: 778–788.

"Public Law 93–473." 1974. http://www.gpo.gov/fdsys/pkg/STATUTE-88/pdf/STATUTE-88-Pg1431.pdf. Accessed on March 11, 2015.

"Public Law 95–617." 1978. http://www.gpo.gov/fdsys/pkg/STATUTE-92/pdf/STATUTE-92-Pg3117.pdf. Accessed on March 11, 2015.

"Public Law 95–618." 1978. http://uscode.house.gov/statutes/pl/95/618.pdf. Accessed on July 24, 2015.

"Public Opinion on Global Issues." 2012. Council on Foreign Relations. http://www.cfr.org/thinktank/iigg/pop/. (The Environment chapter). Accessed on March 18, 2015.

"Renewable Energy Production Tax Credit (PTC)." 2015. U.S. Department of Energy. http://energy.gov/savings/ renewable-electricity-production-tax-credit-ptc. Accessed on March 11, 2015.

"Renewables 2014. Global Status Report." 2015. Renewable Energy Policy Network for the 21st Century [REN21]. http://www.ren21.net/portals/0/documents/resources/ gsr/2014/gsr2014_full%20report_low%20res.pdf. Accessed on March 2, 2015.

Roessner, J. David. 1980. "Implementing State Solar Financial Incentives and R&D Programs." Solar Energy Research Institute. http://www.osti.gov/scitech/servlets/ purl/5211142. Accessed on March 12, 2015.

Sarzynski, Andrea. 2009. "State Policy Experimentation with Financial Incentives for Solar Energy." Institute of Public Policy. George Washington University. http://www2.gwu. edu/~gwipp/GWIPP_Incentive_Inventory.pdf. Accessed on March 12, 2015.

"Selenium Compounds." 2013. U.S. Environmental Protection Agency. http://www.epa.gov/airtoxics/hlthef/ selenium.html. Accessed on March 9, 2015.

Sherlock, Molly F. 2011. "Energy Tax Policy: Historical Perspectives on and Current Status of Energy Tax Expenditures." Congressional Research Service. http://www.leahy.senate.gov/imo/media/ doc/R41227EnergyLegReport.pdf. Accessed on March 11, 2015.

Sherlock, Molly F. 2014. "The Renewable Electricity Production Tax Credit: In Brief." Congressional Research Service. http://nationalaglawcenter.org/ wp-content/uploads/assets/crs/R43453.pdf. Accessed on March 11, 2015.

Soble, Jonathan. 2015. "Short-Circuiting a Solar Boom in Japan." *New York Times*. March 4, 2015. B1+.

"Solar and Wind Energy Credit." 2014. DSIRE. NC Clean Energy Technology Center. http://programs.dsireusa.org/system/program/detail/49. Accessed on March 12, 2015.

"Solar Energy Basics." 2014. National Renewable Energy Laboratory. http://www.nrel.gov/learning/re_solar.html. Accessed on March 11, 2015.

"Solar Energy Operations Impacts." 2015. Tribal Energy and Environmental Information Clearinghouse. http://teeic.indianaffairs.gov/er/solar/impact/op/. Accessed on March 10, 2015.

"The Solar Investment Tax Credit." 2015. Solar Energy Industries Association. http://www.seia.org/sites/default/files/ITC%20101%20Fact%20Sheet%20-%201-27-15.pdf. Accessed on March 11, 2015.

"Solar Power on Landfills." 2015. PV Navigator. http://www.pvnavigator.com/solar-on-lf.htm. Accessed on March 10, 2015.

"Solar Resources." 2014. The Future of Energy. http://energyfuture.wikidot.com/solar-resources. Accessed on March 4, 2015.

"Solar Roadways." 2015. http://www.solarroadways.com/intro.shtml. Accessed on March 10, 2015.

"Solar Voltaic Roadmap." 2010. International Energy Agency. https://www.iea.org/media/freepublications/technologyroadmaps/solar/Solarpv_roadmap_foldout_2010.pdf. Accessed on March 12, 2015.

"Spain Accounts for 72% of the World's Concentrating Solar Power." 2012. Abengoa. http://blog.abengoa.com/blog/2012/08/20/spain-accounts-for-72-of-the-world%E2%80%99s-concentrating-solar-power/. Accessed on March 2, 2015.

"SunShot Initiative." 2014. U.S. Department of Energy. http://energy.gov/sites/prod/files/2014/08/

f18/2014_SunShot_Initiative_Portfolio8.13.14.pdf. Accessed on March 13, 2015.

"SunShot Initiative." 2015. U.S. Department of Energy. http://energy.gov/eere/sunshot/sunshot-initiative. Accessed on March 13, 2015.

"Symptoms, Diagnosis and Treatment." 2015. American Lung Association. http://www.lung.org/lung-disease/silicosis/symptoms-diagnosis.html. Accessed on March 6, 2015.

Taylor, James. 2013. "Florida Gov. Rick Scott Wants Proof Renewable Energy Law Benefits Economy." *Heartland*. http://news.heartland.org/newspaper-article/2013/01/17/florida-gov-rick-scott-wants-proof-renewable-energy-law-benefits-econom. Accessed on March 13, 2015.

"Toxic Land Generates Solar Energy." 2015. National Geographic. http://video.nationalgeographic.com/video/news/us-brownfields-solar-energy-vin. Accessed on March 10, 2015.

"Toxicological Profile for Cadmium." 2012. Agency for Toxic Substances and Disease Registry. http://www.atsdr.cdc.gov/toxprofiles/tp5.pdf. Accessed on March 6, 2015.

Tsoutsosa, Theocharis, Niki Frantzeskakib, and Vassilis Gekas. 2005. "Environmental Impacts from the Solar Energy Technologies." *Energy Policy*. 33: 289–296.

"2.10.2 Direct Global Warming Potentials." 2007. IPCC Fourth Assessment Report: Climate Change 2007. http://www.ipcc.ch/publications_and_data/ar4/wg1/en/ch2s2-10-2.html. Accessed on March 6, 2015.

"2014 Solar Scorecard." 2015. Silicon Valley Toxics Coalition. http://www.solarscorecard.com/2014/2014-SVTC-Solar-Scorecard.pdf. Accessed on March 9, 2015.

"2030 Market Outlook." 2014. Bloomberg New Energy Finance. http://bnef.folioshack.com/document/v71ve0nkrs8e0. Accessed on March 17, 2015.

"Updated Capital Cost Estimates for Utility Scale Electricity Generating Plants." 2013. U.S. Energy Information Administration. http://www.eia.gov/forecasts/capitalcost/pdf/updated_capcost.pdf. Accessed on March 12, 2015.

Vorrath, Sophie. 2015. "Energy Storage to Reach Cost 'Holy Grail,' Mass Adoption in 5 Years." Renew Economy. http://reneweconomy.com.au/2015/energy-storage-to-reach-cost-holy-grail-mass-adoption-in-5-years-18383. Accessed on March 12, 2015.

Wang, Brian. 2011. "Deaths per TWH by Energy Source." Next Big Future. http://nextbigfuture.com/2011/03/deaths-per-twh-by-energy-source.html. Accessed on March 5, 2015.

"The World's 20 Largest Solar Projects." 2015. *Solar Power Today*. http://www.solarpowertoday.com.au/blog/roundup-the-worlds-20-largest-solar-projects/. Accessed on March 13, 2015.

Yassin, Abdiazia, Francis Yebesi, and Rex Tingle. 2005. "Occupational Exposure to Crystalline Silica Dust in the United States, 1988–2003." *Environmental Health Perspectives*. 113(3): 255–260.

3   Perspectives

Introduction

This chapter provides an opportunity for individuals to express their views on very specific aspect of the solar energy issue. These essays may take a position in favor of or opposed to some specific point about solar energy, provide information on technological features, or discuss some other important or interesting concepts involving solar energy.

Distributed and Utility-Scale Solar Power: Brian Angliss

There are two types of solar power. The first type, known as distributed solar power, is solar power that gets installed on the roofs of homes and businesses (also known as "end users"). It's called distributed solar power because the solar energy is used by the home or business where the solar panels are installed. The second type, known as utility-scale solar power, is solar power that is generated by large solar "farms" run by power companies who resell it to end users that may be hundreds of miles away.

Both types of solar power are becoming more common as communities and nations move away from generating

A Nissan Leaf electric car backs out of a charging station at the unveiling of a solar-powered electric car charging station in Portland, Oregon, in 2011. The grid-tied solar canopy has two charging stations with the capacity to fully charge six electric vehicles a day. (AP Photo/Don Ryan)

electricity by burning fossil fuels such as coal and natural gas. While neither type of solar power is perfect, the weaknesses of distributed solar are largely offset by the strengths of utility-scale solar, and vice versa.

Distributed solar power can generate electricity using photovoltaic panels, heat water by passing it through hot solar thermal panels, or do both in what is known as a hybrid photovoltaic/thermal panel. All three types of panels have similar strengths and weaknesses.

The main strength of distributed solar power is that the energy is consumed very close to the point at which it is generated. The solar energy is used by the home or business where the panels are installed, or by neighboring buildings. This saves energy that would otherwise be wasted in between a power plant and the end user.

Distributed solar power has other strengths as well. Distributed systems are easy to design and install on homes and businesses. Distributed systems are also easy to scale to the energy needs of the building on which they are installed; just add or remove solar panels as needed. And since the distributed solar system is installed on an existing building, there are no new environmental concerns with the installation.

The main weakness of distributed solar power is that the Sun doesn't always shine. When the Sun isn't out, the end user must get their electricity from a utility power plant or from batteries. Unfortunately, batteries are expensive to install and maintain, so it is usually cheaper to buy electricity from the power company when the Sun isn't shining.

Another weakness is that distributed solar systems are not very efficient at converting sunlight into electricity. This means that the energy they save by being installed on the home or business is offset by the inefficiency of their energy conversion. Distributed solar is also more expensive to install than a utility-scale system because each end user has his or her own unique needs that prevent solar installers from designing one system that works for everyone. Finally, some neighborhoods or cities have strict rules that limit where solar panels can be installed.

As with distributed solar, utility-scale solar uses a couple of different technologies. Concentrating solar power usually uses lots of mirrors to focus light onto a tower or pipe filled with melted salt that gets even hotter in the sun. The molten salt is used to boil water, and the steam spins the turbine that generates electricity. Utility-scale photovoltaic systems, on the other hand, are large fields of solar panels that are very similar to those installed on a home or business. In this case, however, the photovoltaic system can cover several square miles.

The main strength of utility-scale solar is that it is cheaper to operate. When building a utility-scale solar farm, solar panels or mirrors are manufactured in large numbers, installed on common frames, installed in a single location, and require only one set of building permits. This means that the price per kilowatt-hour (kWh) of electricity is much lower than installing the same number of solar panels on hundreds or thousands of roofs.

Another strength of utility-scale solar is that concentrated solar power technology can be combined with a traditional power plant that burns fossil fuels. Fossil fuel power plants already turn heat into electricity, and so a concentrating solar farm can use the same turbines as the fossil fuel power plant, thus saving a lot of money. Solar farms also tend to have higher conversion efficiency than distributed systems because panels can be programmed to track the Sun instead of being rigidly fastened to a roof. And, finally, concentrated solar power systems can operate for a while after the Sun has stopped shining by storing the molten salt in a huge insulated tank and using the stored heat to generate electricity after the Sun has set.

The main weakness of utility-scale solar farms is their effect on the environment. Utility-scale solar uses land that might be usable for agriculture or left wild. In addition, solar farms are more likely to harm animals and plants that live near the farm, such as by killing birds that may fly through concentrated sunlight.

Another weakness of utility-scale solar is known as the NIMBY (Not in My Back Yard) effect. This term refers to the process that occurs when individuals, businesses, or communities fight against the construction of a utility-scale solar farm because they are afraid that the project will harm them in some way. Utility-scale solar that uses photovoltaic panels has the same battery storage weakness that distributed solar systems do. Finally, utility-scale solar systems waste energy in the electric wires because there can be tens or even hundreds of miles of wire between a utility-scale solar farm and the end users that use its electricity.

Given their strengths and weaknesses, both types of solar power have their place. Utility-scale concentrated solar power makes the most sense at existing fossil fuel power plants because the plant will burn less fuel when the Sun is shining. Utility-scale photovoltaic solar farms make a lot of sense where there is ample undeveloped land. And distributed solar makes sense on new and existing homes and businesses. And when used together, both types of solar power can help meet the world's energy needs.

Brian Angliss is the science editor and climate/energy writer for the website Scholarsandrogues.com and is an electrical engineer working in the Denver, Colorado, metropolitan area. Brian has been a member of the Society of Environmental Journalists since March 2008.

The Value of Solar-Powered Cars: Steven Antalics

Every two years, over one hundred students at the University of Michigan embark on a quest that will take them literally around the world in pursuit of a world championship—though not in a sport that you've likely ever seen broadcast on television.

The sport is known as solar racing, and it pits teams from prestigious universities around the globe in a long-distance road race across North America and the Outback of Australia.

The challenge is a straightforward one: build a street-legal car powered entirely by sunlight, and then see who can cross the finish line first.

Ever since the very first World Solar Challenge (WSC) was organized in Australia in 1987, teams have been building and racing cars that not only showcase the huge advances in solar-cell technologies over the past decades but also the innovative spirit required to change the way we think about energy.

General Motors won that inaugural WSC race and later brought the sport to North America with the first-ever Sunrayce in 1990. The University of Michigan Solar Car Team (UMSolar) won that first American race and later placed third in the WSC in the same year.

That successful beginning laid the foundation for what has become the most successful American solar racing team ever. UMSolar has won a record-setting eight national championships in North America, including winning the last five in a row, and has placed third in the Australian WSC five times. They also won the inaugural Abu Dhabi Solar Challenge in 2015.

These road races are designed not just to highlight the high performance that solar-powered vehicles can achieve; they are also tests of endurance. The World Solar Challenge crosses over 1,800 miles (3,000 km) of the harsh Australian Outback, stretching all the way from Darwin to Adelaide. The American Solar Challenge route has varied over time, but its longest distance was a 2,500-mile (4,000-km) odyssey from Austin, Texas, to Calgary, Alberta.

In the 1990s, when solar cells were significantly less efficient than they are today, the challenge was to build a car that could run a long distance at normal highway speeds. In recent years, advances in solar-cell efficiencies have helped solar cars maintain speeds of more than 60 mph (97 km/h) over the course of the entire race, relying on battery packs to power through areas of clouds or even rain.

The top speed of UMSolar's cars is now over 90 mph (145 km/h).

Given these high speeds, racing regulations have changed from emphasizing speed to practicality. Drivers are now made to sit upright, instead of leaning back, and the WSC has even created a new "Cruiser" class for solar-powered cars seating two to four people.

How have solar-powered cars, which are limited to using solar arrays of approximately 2 square meters (21 sq ft), been able to achieve such speeds on such little energy?

Let's start by putting the available energy in perspective. The amount of sunlight falling on the Earth's surface at any given time is known as solar irradiance, and it averages a nearly constant 1,367 watts per square meter (World Energy Council 2013). A two-square-meter solar array is able, in theory, to capture a maximum of approximately 2,700 W if 100 percent of the Sun's energy were transferred to electricity and you were at the Equator. That's about the same power as a typical hairdryer, but solar cars have to make do with significantly less.

First, there is less Sun power available the farther away you are from the Equator. Second, there is no method of energy transfer that is 100 percent efficient, and solar cells are no exception.

Solar cells typically come in two "flavors:" those made from silicon crystals (Si) and those made from gallium arsenide (GaAs) crystals. Silicon cells are typically cheaper but less efficient—only about 25 percent of the sunlight is converted into electricity. GaAs cells, which are much more expensive, are also much more efficient and are able to convert upward of 45 percent of the Sun's energy directly into electricity ("Best Research-Cell Efficiencies" 2014).

Solar car teams use both types of cells, but UMSolar and most of the top teams use GaAs cells to optimize their power. Teams have also employed other technologies to help increase the power of their solar-cell arrays, including high-tech coatings and encapsulation (to prevent damage to the cells and also increase their efficiency), cell "shingling" (where corners of cells are overlapped, like shingles on a roof), and concentrators, which use small mirror systems to increase the amount of light shining on the cells (Gochermann Solar Technology 2015).

However, having the most powerful solar array is no guarantee that your solar car will finish first. There are numerous other challenges to overcome when designing a solar racer.

We tend not to think too much about it when driving modern cars, but the single biggest force to overcome at highway speeds is air resistance, or aerodynamic drag. Even though cars with internal-combustion engines have significantly more power available to overcome air resistance, automotive companies still work to minimize drag because that helps maximize a car's fuel economy and performance ("Drag Queens: Aerodynamics Compared" 2014).

The fastest solar cars, even though they're street-legal, are more than twice as aerodynamic as the best production cars. They achieve such aerodynamic performance by using lightweight materials, innovative technology, and advanced body designs.

In order to reduce the weight of their cars, which are less than 500 lb (225 kg), UMSolar creates custom parts from lightweight aluminum, titanium, and carbon fiber. These same materials are also used in the high-end race cars of Formula One and IndyCar.

UMSolar's cars have also used rear-facing cameras to avoid having to mount side-view mirrors on the car, which would decrease the car's aerodynamics. Such cameras are now typically found in many consumer vehicles to avoid obstacles and help drivers park.

Since solar cars are only designed for racing, they were made to be as flat and thin as possible, while still being safe enough to protect the driver. In recent years, regulations have changed so that the driver now has to sit fully upright (as in a normal car) and use a full-size steering wheel. All of this increases the air resistance on the car and forces ever-more-innovative designs to compensate.

One such design feature is the concept of "windowed" fairings. The spokes in a typical wheel "catch" air and increase drag, and the only way to reduce that drag is to either use a solid wheel or to cover the wheel with the car's lower body, or "fairing." Solid wheels, often made of carbon fiber, have one serious drawback: they are fragile. When you're racing for thousands of miles, you don't want to lose time changing the rims if they break. Therefore, most teams, including UMSolar, choose to use a traditional spoked rim and cover the wheel.

Covering the wheel presents a unique problem. The narrower you make the fairing, the better the aerodynamics, but if you make the fairing too narrow, the driver won't be able to turn the wheels.

Teams have tried several designs to solve this problem, including using four-wheel steering so that each wheel has to turn less. In the end, the best solution has been to make the wheel-covering fairings very narrow but to cut a "window" that opens and closes automatically when the driver turns the steering wheel.

The final element of building a championship-winning solar car team is to learn how to excel as members of a high-performing team. This is perhaps the hardest lesson in solar racing, but the one that is the most rewarding in the long run.

Windowed fairings are a good example of teamwork. Students from different engineering disciplines all have competing ideas of the "best" solution. In the end, team members have to learn how to communicate their ideas clearly to one another and also how to compromise in pursuit of a shared goal. If they cannot decide, the car won't be built, and it'll never race.

Communication is key during the races as well. The best teams have dedicated groups that go out and scout the route before and during the races, reporting on everything from obstacles to weather to roadkill. The car itself also has wireless connectivity with the rest of the race team so that its performance can be modified and adapted to fit the conditions.

The experience of designing, funding, building, and racing a solar-powered car across two continents is a transformative one for the members of UMSolar. These projects not only provide real-world research and testing for the most cutting-edge automotive and solar-powered technologies, but they also teach students the skills they'll need to change the way we think about and use energy in our everyday lives.

References

"Best Research-Cell Efficiencies." 2014. National Renewable Energy Laboratory. http://www.nrel.gov/ncpv/images/ efficiency_chart.jpg. Accessed on May 2, 2015.

Bridgestone World Solar Challenge. 2015. http://www .worldsolarchallenge.org/. Accessed on May 5, 2015.

"Drag Queens: Aerodynamics Compared." June 2014. Car and Driver. http://www.caranddriver.com/features/ drag-queens-aerodynamics-compared-comparison-test. Accessed on May 5, 2015.

"Features." 2015. Gochermann Solar Technology. http:// www.gochermann.com/array_features/. Accessed on May 5, 2015.

"University of Michigan Solar Car Team—History" 2015. University of Michigan Solar Car Team. http://www .solarcar.engin.umich.edu/. Accessed on May 5, 2015.

World Energy Council. 2013. "World Energy Resources Survey." http://www.worldenergy.org/wp-content/ uploads/2013/09/Complete_WER_2013_Survey.pdf. Accessed on May 5, 2015.

Steven Antalics is a former leader of the University of Michigan Solar Car Team. A lifelong enthusiast for renewable energy and technological innovation, he currently works as a business analyst for a global commodities trading company.

Solar Energy—Thermal versus PV: Jim Conca

Numbers matter. When planning for our nation's future, we need to choose a mix of energy sources that will provide the energy we need without destroying too much of our planet's ecosystem.

For energy production, these important numbers include:

- the cost to produce a kilowatt-hour (kWh) of electricity,
- the death print (the number of people who die per kWh of electricity produced),
- the carbon footprint (the amount of carbon emitted per kWh of electricity produced),
- the physical footprint (the area required to produce a kWh of electricity), and
- the capacity factor, or cf (the percentage of power produced relative to maximum possible).

These last three numbers are important to understand when comparing the two main types of solar energy systems—solar photovoltaic (PV) and solar thermal (also called *concentrated solar*).

PV cells transform incident solar radiation directly into electricity. Solar thermal uses long rows of computerized parabolic mirrors to track the Sun and reflect its rays to boilers atop giant towers, or to concentrate sunlight on a tube filled with a heat transfer fluid. The heat is used to create superheated steam, which spins a conventional steam turbine to generate electricity just like an ordinary coal, gas, or nuclear plant.

The carbon footprints for each type of solar are about the same, between 50 and 100 gCO_2/kWh produced, and mostly involve manufacturing of the materials and construction at the site. Solar thermal often needs some help from a natural gas plant to get started in the morning, somewhat increasing its footprint.

The physical footprints can be quite different. Solar energy requires a fair amount of space to absorb sufficient sunlight to contribute a significant amount of energy. Solar PV can be placed on rooftops, the tops of parking structures and buildings, and on many other existing structures, so the effective physical footprint is smaller since you don't have to take up new space. My own 4 kWh array on my roof did not require any additional space.

But solar thermal requires new space and that may impact ecological habitat. We can place these facilities in the desert without destroying too much fragile ecosystems by building them on land whose ecosystem is already destroyed beyond the point where it can repair itself. Many of these degraded areas have been mapped by organizations such as the Nature Conservancy, and that has limited the amount of habitat impact from large solar arrays.

The third important number is the capacity factor. For solar, it depends strongly on where the systems are, what the weather is like, how often the Sun shines without clouds or storms, and how well the cells or mirrors are positioned relative to the Sun's rays. The capacity factor (cf) for solar is generally about 20% but should be about 30% in desert areas like the southeastern United States, where the Sun shines without cloud cover for over 300 days a year.

When any energy system is built, it has something called a nameplate installed capacity, which is the maximum power produced when everything is running perfectly. The capacity factor is equal to what the power plant produces per year divided by what it could produce if it ran at capacity, 24 hours a day, every day for the entire year.

However, no power plant runs all the time. There are outages for refueling, maintenance, and accidents. Often the Sun isn't shining or the wind isn't blowing. The capacity factor for wind is generally 30 percent, or 0.30, although it goes from 20 percent in some places in the Pacific Northwest to 40 percent in Tornado Alley that runs up from central Texas to the Dakotas. Nuclear is generally about 90 percent, or 0.90, everywhere.

The largest solar energy facility in the world completed its first full year in 2014 and shows just how important the capacity factor is. The Ivanpah Solar Electric Generating System is a 5-square-mile, 392 MW nameplate thermal solar array that cost $2.2 billion. It is located in the Mojave Desert and is supposed to produce a billion kilowatt-hours per year for the next 25 years. It relies on 347,000 sun-tracking mirrors that direct highly concentrated solar radiation at boilers sitting atop three 450-foot-tall towers.

But there's a problem: the Sun does not seem to be cooperating. A spokesman for the owner, NRG Energy, blamed the weather, saying the Sun didn't perform as predicted!

Ivanpah is supposed to produce enough power for 140,000 homes, or 1 billion kWh, per year; 392 MW translates to 3.4 billion kWh per year if it produced at capacity all the time:

392 MW × 1000 kW/MW × 8766 hours/year = 3.4 billion kWh

So 1 billion kWh equates to a capacity factor (cf) of about 30 percent:

1 billion kWh ÷ 3.4 billion kWh = 0.29 = 29 percent for its cf.

However, in 2014, Ivanpah produced less than 0.4 billion kWh of electricity:

0.4 billion kWh ÷ 3.4 billion kWh = 0.117 = 11.7 percent for its cf.

This is a very low cf, much lower than for most solar PV systems.

And then there is the water needed for the project. Ivanpah uses about 32 million gallons of groundwater each year to keep its boilers full and mirrors clean. In a desert, that's not chump change.

"The Mojave looks dry on the surface, but there's ancient groundwater deep underground," says Sophie Parker, a California-based regional ecologist for the Conservancy. "People are pumping it out much faster than it can be naturally replenished. It's basically fossil water. Like fossil fuels, we'll eventually just run out" (Paskus 2015).

Until recently, PV cells were expensive enough that it was thought these thermal solar arrays would be the best way to make solar feasible on a large scale. But PV cell costs have come down in recent years, and their efficiencies have gone up. PV can be emplaced at any size and distributed over existing surfaces, like rooftops, directly in areas that need the power.

Thus, you do not need to cover miles of desert in glass and metal, don't need water to operate, don't need turbines, and don't need to conduct the electricity great distances.

Just ask Warren Buffet. He recently purchased the world's largest photovoltaic solar array further north in Bakersfield, California. It's also a 5-square-mile array, but it is a 579 MW nameplate PV solar, about 50 percent more powerful than Ivanpah, and has a capacity factor of over 20 percent. Buffet paid about the same price as Ivanpah.

So PV solar will undoubtedly become the world's primary source of solar power. Just read the numbers.

Reference

Paskus, Laura. 2015. *Nature Conservancy Magazine.* http://magazine.nature.org/features/place-in-the-sun.xml. Accessed on May 1, 2015.

Recommended Links

http://www.brightsourceenergy.com/ivanpah-solar-project#.VDB8pSjx—k

http://www.businessinsider.com/ivanpah-solar-plant-already-irrelevant-2014-2

http://cleantechnica.com/2014/06/11/aus-concentrating-
solar-power-breakthrough-hit-us-shores/

http://www.forbes.com/sites/jamesconca/2014/09/08/saudi-
arabia-fast-tracks-nuclear-power/

http://www.greenbiz.com/blog/2014/02/19/largest-solar-
thermal-plant-completed-ivanpah

http://magazine.nature.org/features/place-in-the-sun.xml

http://www.renewableenergyworld.com/rea/news/article/
2009/10/solar-electric-facility-o-m-now-comes-the-
hard-part

*Jim Conca is a PhD geologist, researcher, author, and blog-
ger for Forbes.com who writes extensively on energy issues and
sustainability.*

The SolaRoad Project: Gina Hagler

Traditionally, sources of solar power have been dedicated to the
single purpose of turning energy from the Sun into electricity. From solar panels in open fields and other sunlit areas to
solar panels atop houses, these photovoltaic systems existed for
this sole purpose. While it is true such installations could be
optimized for the greatest efficiency, there were still significant
limitations. The systems were expensive, and since they used
the space they occupied for a dedicated purpose, they were limited in utility and required the creation of a costly asset.

A new generation of sources that generate electricity from the
Sun is now in development. These sources generate solar power
as a by-product of another function. They are not currently as
efficient as dedicated systems, but since they are serving a dual
purpose with another structure, any energy generated is energy
that would not have been generated in the past. One such type
of system is a roof for an existing parking structure. The roof
provides shade for the cars while generating energy in an area
that already had a footprint. If the roof can be used to channel

rainwater so as to minimize the effect of runoff on bodies of water impacted by that runoff, further reducing the impact on the environment, so much the better.

In the Netherlands, where bicycles constitute a significant form of transportation, an ambitious and innovative use of solar technology is currently in testing and development by TNO, the Netherlands Organisation for Applied Scientific Research. Their work in applied science and interest in renewable energy led them to form a consortium with partners who had expertise in areas such as marketing. The consortium explored the feasibility of using solar panel materials in bike paths and roadways to generate electricity as a by-product. Their research led to the creation of a solar bike path prototype in Krommenie—an area in the province of North Holland with a deep interest in renewable energy and environmental conservation—with future plans to expand the program to bike paths and roadways elsewhere in the Netherlands and beyond.

A 100-meter bike path was constructed as a prototype because the path could be positioned to allow easy access to the components that needed to be monitored for durability and reliability. The manageable, yet significant, length of the path and the ability to record performance of the surface under a variety of weather conditions was another consideration. TNO knew the materials had to be durable enough to withstand bicycle traffic—and service vehicles as well—but because it was a bike path, it was not necessary that the components stand up to the constant travel of heavy vehicles such as those that would be the primary mode of transportation on a roadway. With the bike path, the materials could be tested and it could be ascertained that the characteristics of an efficient solar panel could coexist with the characteristics that give the travel surface sufficient traction before stepping up the requirement of heavier modes of transport.

In the prototype, standard silicon solar cells were used. (Future plans call for the use of other types of solar cells, but a specialized solar cell is not required.) Since a glass surface is part of the cells, there was a concern about what would happen if the glass in the panels were to break. Because of this concern, the glass is shatterproof to guard against injury.

The roadbed for the bike path was constructed so that the panels of the path could be connected without causing seams that would make it untenable to ride comfortably on the path. The road also needed to be able to withstand damage from the soil or from tree roots or other impediments. The output for the first bike path was estimated to generate enough electricity for its own needs, as well as several households' annual electricity needs.

The SolaRoad bike path in Krommenie opened in 2014. It can generate sufficient electricity to power lights for the path at night at the very least, and ideally traffic signals or more in the future. To date, the path has been a great success. It lights itself within the roadway when the Sun is down and generates the electricity for that purpose while the Sun is up.

In the Netherlands, a bike path makes a lot of sense. Bicycles are a common form of transportation, and experts estimate that many roadways will not be ideal candidates for solar technology because their orientation to the Sun is not well suited. In countries where the automobile is the primary form of transportation and roadways and highways cover a larger total area of the land surface, it will be feasible to generate far more electricity from the roads than from bike paths.

The current model for SolaRoad going forward is for the investment in the roadway to be recouped over a 20-year period. The ultimate goal is for it to be recouped within 15 years. Where toll roads are in use or trucks pay fees to use the roadways, it is not inconceivable that the investment can be recovered more quickly. The potential for renewable energy from a source that has a dual purpose is one that has not existed at this scale in the past. With energy brought into the existing grid—or funneled into private grids—through solar roadways such as SolaRoad, it is plausible that dependence on traditionally generated electricity can be reduced by a source close to intended use and without the necessity for a large investment in infrastructure.

TNO has changed the paradigm. Rather than look for locations where dedicated solar energy installations can be constructed as has been the case in the past, communities and transportation authorities now have not only a new way to look at solar energy installations—as secondary purpose to an existing or planned bike path or roadway—but also a successful proof of concept of dual-purpose installations to use as the baseline.

An image of the SolaRoad concept is available online at http://www.solaroad.nl/en/hoe-is-het-idee-ontstaan/.

Gina Hagler is a science and technology writer. Her work has appeared in numerous publications as well as regularly on Bitter-Empire Sci-Tech. Modeling Ships and Space Craft was published by Springer Verlag in 2012. You can see more of her work at gina-hagler.com, where she blogs on regularly.

Solar Financing: Adam Johnston

In recent decades, society has been using Sun power more in improving livelihoods, and creating a more sustainable lifestyle for the planet. The International Renewable Energy Agency notes sharp declines in residential photovoltaic (PV) (75 percent since 2009) and utility-scale solar costs (29 percent to 65 percent from 2010 to 2014, based on regional variances) because improved technologies are now helping solar energy become mainstream, making it more attractive to consumers and investors ("Renewable Power Generation Costs" in 2014 2015).

Declining solar energy prices have helped boost booth capacity and financing. Overall global solar PV capacity reached 177 GW in 2014 (compared to 0.105 GW in 1992), thanks to 37.4 GW of new installations ("Renewable Power Generation Costs" 2015).

Meanwhile, global solar financing for panel installers and manufacturers increased 175 percent in 2014, reaching $26.5 billion, compared to $9.6 billion in 2013 (Mercom Capital Group 2015).

Concerns over climate change, increased global population, and rise of the middle class from developing nations (Friedman 2008) will only further the importance of improving accessibility to financing solar projects.

Solar installers require funding to help cover the costs of furnishing rooftop panels, while looking to expand the industry by making solar energy more accessible to more income levels. This is known as third-party financing, and makes up for more than half of funding new U.S. installations, in Arizona, California, Massachusetts, and Colorado, according to Greentech Media (2013). By 2016, the U.S. home solar financing market is expected to reach $5.7 billion, four times more than in 2012.

This model allows for businesses and residential customers to sign a deal with a third-party provider for electricity created by a third-party solar system, which is operated, maintained, and run by the company. Investors take advantage of the federal 30 percent solar investment tax credit (ITC) for home and business solar installations, while the installer receives regular payments from the customer.

One recent example is Google's SolarCity, providing $300 million in tax equity financing for new home solar installations (Martin 2015). Google will be able to leverage the tax breaks in its favor, as unprofitable companies, like SolarCity, cannot benefit from the ITC.

Consumers have two major options in third-party solar financing: leasing and power purchase agreement (PPAs). A leasing agreement allows consumers to buy service of a solar system's electrical generation, without buying an actual system. PPAs provide customers an opportunity to buy a solar system's electrical generation, without buying the system parts, creating the solar electricity ("Third Party Solar Financing" 2015).

Leasing a solar system provides consumer benefits that PPAs do not have. Homeowners do not have to pay up-front costs, while not worrying about concerns regarding the maintenance of solar panels (Fuscaldo 2014).

Leasing has helped U.S. solar home growth. In 2014, residential installations reached 1 GW for the first time, and 50 percent growth for the third consecutive year (Litvak 2015). SolarCity (46 percent) and Vivint Solar (34 percent) captured 80 percent of all U.S. PV residential installations in 2014.

While third-party financing has helped boost solar installations, and making this renewable resource more accessible, crowdfunding may provide further assistance in putting solar in the hands of low-income communities, and push financial costs further down.

Crowdfunding uses the power of web-based technologies, allowing people to pool money to start new enterprises. Examples of crowdfunding websites include Kiva, Indiegogo, and CrowdRise.

Known as the "kick-starter" of solar crowdfunding, Oakland, California-based, Mosaic has leveraged this financing method to allow average people to invest in solar projects, while improving access to solar energy.

An investor may put down as little as $25 for various projects. Billy Parish, founder and president of Mosaic says, "Crowdsourcing solar can create good investment opportunities that also help decentralize our financial system and enable more people to benefit from the transition to a clean energy economy" (Frishberg 2013).

Mosaic's initial crowdfunding campaign in January 2013 raised $313,000, enough to finance four California-affordable solar energy projects. This campaign saw an average investment of $670 per each person (Gilpin 2014). Other campaigns have helped provide solar power to military residencies and schools. Investors can get between 4.5 and 7 percent returns, with borrowers paying back the loan in 10 years.

As solar crowdfunding can provide more access to solar financing, it also can drive borrowing costs further down.

Developers and tax equity financiers are the two main sources of capital (interest rates at 17 and 18 percent). This keeps capital costs high, spilling into consumer electricity solar prices, suggests Jesse Morris of the Rocky Mountain Institute (2013).

However, Morris suggests when crowdsourcing is added, the overall capital interest rate falls, while the price of solar electricity also declines. Crowd financing has a lower rate of interest (6.5 percent), compared to tax equity and developer financiers. If all funding came from crowd sources, that would provide very low electricity and interest rate costs, based on Morris's analysis.

Providing alternative funding sources is critical in advancing solar capacity. The International Energy Agency (2014) projects solar capacity to reach over 400 GW by 2020, and as high as 4,674 GW by 2050. Adding crowdfunding to finance solar finance solar projects, helping to both drop costs and improve access to those who can't afford financing, is one trend to watch in the future.

Improved technology has helped push solar costs down, making it more attractive to consumers and investors. Declining costs have advanced both solar capacity and solar financing needs. Climate change concerns and a rising middle class from developing nations also reinforce the need for accessibility and lowering financing costs to increase solar capacity.

Solar installers need financing in order to cover their costs and allow consumers the opportunity to use solar as a clean energy source. Third-party financing is the most well-known option, as investors provide financing to downstream companies (installers), who then get the benefit of the 30 percent ITC, while installation companies receive revenue from consumers using their service. Two third-party options consumers use for installing solar are leasing and PPA, with leasing being the more popular option. This pushed 1 GW of U.S. residential installations for the first time, in 2014.

As analysis projects solar capacity to reach over 4,600 GW by 2050, crowdfunding may provide a critical component in financing, given its ability to democratize finance, while lowering project costs further.

References

Friedman, Thomas. 2008. *Hot, Flat, & Crowded: Why We Need A Green Revolution—And How It Can Renew America.* New York: Farrar, Straus and Giroux.

Frishberg, Manny. 2013. "Alternative Financing for US Solar Energy." *Research Technology Management.* 56(6): 7–8.

Fuscaldo, Donna. 2014. "Buy vs. Lease: Solar Panels on Your Home." Fox Business. http://www.foxbusiness.com/personal-finance/2014/01/29/buy-vs-lease-solar-panels-on-your-home/. Accessed on April 20, 2015.

Gilpin, Lyndsey, 2014. "How Crowdfunding Solar Power Is Democratizing the Way We Finance Clean Energy." *TechRepublic.* http://www.techrepublic.com/article/how-crowdfunding-solar-power-is-democratizing-the-way-we-finance-clean-energy/. Accessed on April 18, 2015.

Litvak, Nicole. 2015. "Here Are the Top 5 Residential Solar Installers of 2014." Greentech Media. http://www.greentechmedia.com/articles/read/Here-Are-the-Top-Five-Residential-Solar-Installers-of-2014. Accessed on April 20, 2015.

Martin, Christopher. 2015. "Google Is Making Its Biggest Ever Bet on Renewable Energy." Bloomberg. http://www.bloomberg.com/news/articles/2015–02–26/google-makes-biggest-bet-on-renewables-to-fund-solarcity. Accessed on April 23, 2015.

Mercom Capital Group. 2015 "Total Corporate Funding Increases 175 Percent to $26.5 Billion in the Solar Sector, VC Funding Doubles, Strong Public Market, Debt Financing and IPO Activity." Mercom Capital Group. http://mercomcapital.com/total-corporate-funding-

increases-175-percent-to-$26.5-billion-in-the-solar-sec
tor-vc-funding-doubles-strong-public-market-debt-fin
ancing-and-ipo-activity-reports-mercom-capital-group.
Accessed on April 13, 2015.

Morris. Jesse. 2013. "Rocky Mountain Institute: How
Crowdfunding Lowers The Cost of Solar Energy." Green
Biz. http://www.greenbiz.com/blog/2013/02/27/how-
crowdfunding-lowers-cost-solar-energy. Accessed on April 20,
2015.

"Renewable Power Generation Costs in 2014." 2015.
International Renewable Energy Agency. http://www.irena
.org/DocumentDownloads/Publications/IRENA_RE_
Power_Costs_2014_report.pdf. Accessed on April 13,
2015.

"Technology Roadmap." 2014. International Energy Agency.
https://www.iea.org/publications/freepublications/
publication/TechnologyRoadmapSolarPhotovoltaicEnergy_
2014edition.pdf. Accessed on April 26, 2015.

"2014 Snapshot of Global PV Markets." 2015. International
Energy Agency. http://www.iea-pvps.org/fileadmin/dam/
public/report/technical/PVPS_report_-_A_Snapshot_
of_Global_PV_-_1992–2014.pdf. Accessed on April 13,
2015.

US Department of Energy (DOE). 2015. "Third Party Solar
Financing." Energy.gov. http://apps3.eere.energy.gov/
greenpower/onsite/solar_financing.shtml. Accessed on
April 26, 2015.

"US Residential Solar Financing to Reach $5.7 Billion
by 2016." 2013. Greentech Media. http://www.greentech
media.com/articles/read/us-residential-solar-financing-
to-reach-5.7-billion-by-2016?utm_source=feedburner
&utm_medium=feed&utm_campaign=Feed%3A+GTM_
Solar+%28GTM+Solar%29&utm_content=Google+
Feedfetcher. Accessed on April 20, 2015.

Adam Johnston is a blogger living in Winnipeg, Manitoba, Canada. He has written for numerous cleantech websites, including CleanTechnica and Solar Love. A graduate of the University of Winnipeg with a BA in economics and rhetoric and communications, Adam currently is working on his professional development certificate from the University of Toronto School of Environment program in Renewable Energy. Adam is also planning on starting a part-time sustainable development consulting business, and he owns his own income tax preparation enterprise on the side.

Current Limitations in Solar Energy and Solutions for the Future: Yoo Jung Kim

Despite its potential as a renewable energy source, solar power currently suffers from a number of economic and technical hurdles that have prevented it from becoming more widely adopted by the public. Fortunately, new developments may enable solar energy to displace a significant amount of energy consumption produced by fossil fuels in the near future.

First, although an existing solar cell requires no additional fuel to convert solar energy into electrical energy, for the total benefits of solar energy to outweigh those of fossil fuels, consumers must account for the fact that the manufacturing and the disposal of solar cells require money, consume fuel, and produce waste. This conundrum has been dubbed "Solar Energy's Red Queen Effect," after a quote from Lewis Carroll's *Through the Looking Glass*, in which the Red Queen remarks, "Now here, it takes all the running you can do to keep in the same place" (Schneider 2008). In 2005, Andy Black—the CEO of OnGrid Solar—pointed out the paradox of a clean-energy industry powered by coal energy, stating, "We're not going to make a difference unless we grow fast" (Schneider 2008).

Secondly, solar energy must contend with the supposed limitations of the grid. In solar-energy-rich states like California, Arizona, and Hawaii, the excess electricity generated by solar panels is typically funneled into a power grid. The extra electricity can then be used to power another house or building, and the utility companies pay the owner of the solar system for excess electricity. However, the growing number homeowners installing solar panels has added new strains on existing circuits and power lines—resulting in unexpected voltage fluctuations, circuit overloads, and blackouts—and has cut into the revenue of electrical companies. Both of these factors have already led some utility companies to discourage homeowners from embracing solar energy by removing incentives, delaying the application process for solar energy, and adding extra fees and charges for utility customers with existing solar systems (Cardwell 2015).

Fortunately, with regard to the Red Queen Effect, solar energy has been able to "grow fast" in the last decade, thanks to technological advances and increasing production efficiency, resulting in solar cells that require less energy and money to produce.

For example, in the 2000s, Martin Green, Scientia Professor at the University of New South Wales and a major proponent of solar energy, evaluated how much each physical component of a commonly produced solar cell—including the polycrystalline silicon, the glass cover, and the silver wires—would cost to produce. Green then announced that as long as solar cells were made out of polycrystalline silicon, the price of solar energy would never break below $1 per watt (McKenna 2015). Since then, using low-cost methods for manufacturing polycrystalline silicon, replacing glass covers with plastic, reducing the amount of silver used, and incorporating more automation in the manufacturing process have drastically reduced the cost of solar cells from $4 per watt in 2007 to $0.50 per watt in 2014 (McKenna 2015).

Eventually, we may see the cost of producing solar modules coming down to 25 cents per watt. To reduce this price even further, the energy conversion efficiency for polycrystalline silicon–based cells will have to rise above the current limit of 22.9 percent. According to Green, this may soon be possible with the development of materials that could be placed on top of solar cells to harness energy from a wider range of the electromagnetic spectrum (McKenna 2015). With these advances in technology, grid parity—the point at which renewable energy such as solar and wind power can produce electricity at a cost equal to that of power bought from the electrical grid—may soon be a reality.

Furthermore, making small improvements to existing grids and enabling consumers to go off the grid completely may resolve current inefficiencies faced by utility companies. In 2011, the International Energy Agency found that contrary to popular belief, efficient integration of solar energy would mostly require better planning and modest upgrades by utility companies to account for variability, rather than having overhauls to existing grid systems ("Solar Energy Perspectives" 2011). This finding and others like it have prompted a utility commission to force a utility company in Hawaii to make these small upgrades to serve the local homeowners (Cardwell 2015).

Moreover, improvements in home battery technology may help homeowners bypass the power grid completely. In April 2015, Elon Musk, the CEO of Tesla, unveiled the Tesla Firewall, a compact, wall-mounted, lithium-based battery for homes (Cardwell 2015). With these batteries, a house can run on live solar energy during the day, save the unused energy generated by the panels within the battery, and run on battery-stored power on overcast days and during the night.

In short, solar energy has to overcome real challenges in order to reach grid parity and be fully embraced by the average homeowner living in a Sun-rich state, but recent developments—including more efficiently produced solar cells and the growing availability of home batteries for consumers—indicate a sunny outlook.

References

Cardwell, Diane. 2015. "Solar Power Battles Puts Hawaii at Forefront of Worldwide Changes." The New York Times. http://www.nytimes.com/2015/04/19/business/energy-environment/solar-power-battle-puts-hawaii-

at-forefront-of-worldwide-changes.html. Accessed on
April 30, 2015.

Cardwell, Diane. 2015. "Tesla Ventures into Solar Power
Storage for Home and Business." The New York
Times. http://www.nytimes.com/2015/05/01/business/
energy-environment/with-new-factory-tesla-ventures-
into-solar-power-storage-for-home-and-business.html.
Accessed on May 2, 2015.

McKenna, Phil. 2015. "Solar Power Will Soon Be as Cheap as
Coal." Quartz. http://qz.com/386261/solar-power-will-soon-
be-as-cheap-as-coal/. Accessed on April 25, 2015.

Schneider, David. 2008. "Solar Energy's Red Queen."
American Scientist. http://www.americanscientist
.org/issues/pub/solar-energys-red-queen. Accessed on
April 30, 2015.

"Solar Energy Perspectives." 2011. Paris: Organisation for
Economic Co-operation and Development/International
Energy Agency.

*Yoo Jung Kim is a recent graduate of Dartmouth College. She
is a staff columnist for the Public Library of Science Sci-Ed Blog
and is currently coauthoring a book* titled *What Every College Science Student Should Know,* to be published by the University of
Chicago Press *in 2016.*

Artificial Photosynthesis: David Latchman

It is remarkable that plants have evolved to do over millions of
years what humans are just learning to do. They take energy
from the Sun and use the carbon dioxide in the air around
them to create their cellular structure and the fuel they need
to survive. This process is far from simple and is a complex,
multistage process that, until recently, was a mystery. The clos-
est we have to what plants do is solar energy—specially doped
semiconducting materials that convert the Sun's photons into
moving electrons.

As remarkable as that is, solar power is not without its problems. On average, there is more solar energy striking Earth's surface in one and a half hours than the 2001 worldwide consumption from all sources combined. Unfortunately, sunlight is not constant throughout the day, and we have no way of storing the energy when it is highest or needed most. If we could somehow mimic what plants do and convert sunlight, water, and carbon dioxide into usable fuels, we may be able to overcome some of the limitations of solar power.

Mimicking Photosynthesis

The plant photosynthetic reaction occurs in two half-reactions (oxidation and reduction) to produce sugar. In the first stage, chlorophyll captures sunlight, and a collection of proteins and enzymes photo-oxidize water molecules to release oxygen and protons (hydrogen ions). The second stage, the Calvin-Benson cycle, uses the charge carriers, the electrons and protons, from the previous process to convert carbon dioxide and water into carbohydrates.

Plants have been using sunlight to convert water and carbon dioxide into sugars and cellulose for millions of years but the process is surprisingly inefficient. Plants, at most, convert about 1 percent of the Sun's energy used during photosynthesis into sugars and cellular matter, although some plants, such as sugar cane and some species of algae, have higher efficiencies—about 3 percent. The goal of artificial photosynthesis is to develop materials that mimic or replicate this process to produce fuel more efficiently than plants. This fuel can then be stored and used when sunlight is not available.

The key to achieving artificial photosynthesis is to develop new catalysts that replicate both stages in the photosynthetic reaction—one to oxidize water and the other to absorb and reduce carbon dioxide. When these catalysts are coupled, creating usable hydrocarbon fuels such as methanol from sunlight, water, and carbon dioxide becomes possible.

While it is possible to use electrolysis from photovoltaics to break water into its constituent elements, that is, hydrogen and oxygen, it is a very inefficient process to generate hydrogen fuel. It takes more energy to break water's covalent bonds than the 2.5 volts predicted by theory. This higher-than-theoretically-predicted voltage needed to electrolyze water is known as overpotential, and it depends mostly on the nature of the electrodes and design of the cell.

Water Oxidation

The first step in making artificial photosynthesis possible is to find the right semiconductor photocatalyst to break water into its constituent elements. As light falls on the surfaces of these materials, surface electrons are excited and react with water molecules to split them apart. One example of a photocatalyst is titanium dioxide (TiO_2). First discovered in the late 1960s by Japanese chemist Akira Fujishima, this cheap and abundant material suffers from a fatal flaw that prevents it from becoming a promising photocatalyst.

The titanium dioxide flaw stems from its wide bandgap, which limits light absorption in the ultraviolet range. To make this a better photocatalyst, scientists search for ways to narrow the bandgap into the visible range and, thereby, improve catalytic efficiency. This can be done by doping the material, or adding trace impurities to the crystal structure, to change the material's bandgap. The material's surface can also be prepared and coated with special photoreactive dyes that react with light in the visible range to inject electrons into the semiconductor's surface. These electrons then oxidize and split water molecules.

Titanium dioxide's less than desirable properties have led researchers to look for other materials that are cheap and abundant, and can be used as effective photocatalysts. Some materials currently being investigated are manganese, cobalt, rubidium, iron oxide, and iridium. Manganese is found at the photosynthetic core of plants, and a single atom can trigger the natural processes that use sunlight to split water. Like many of the other materials, though promising, manganese also has its drawbacks. Many of the materials derived from manganese are not terribly stable and degrade in sunlight or water, the very conditions they are expected to perform.

Some scientists are also looking to nature, or at least seeking ways to modify organisms to help replicate the water oxidation part of photosynthesis. Some species of algae, for example, *C. reinhardtii*, when deprived of sulfur, can switch to from oxygen to hydrogen production. In the same way we can use the hydrogen derived from photocatalysts in future reactions to create synthetic fuels, we can genetically modify organisms to do the same. In essence our fuel production cell is living.

Reducing Carbon Dioxide to Create Fuels

The next step in the process is to reduce carbon dioxide using the hydrogen in the previous process to create a carbon-based fuel, preferably methanol. Plants convert inorganic carbon into organic compounds through a process known as carbon fixation by using the enzyme RuBisCO (ribulose-1,5-bisphosphate carboxylase/oxygenase). Artificial CO_2 reduction aims to do the same but at a faster and more economical rate; plants incorporate a few molecules of carbon dioxide a minute on average.

This step of reducing carbon dioxide is an important one as it can be used to create more energy-dense fuels. Like the water oxidation catalysts before, finding the right catalysts to reduce CO_2 that is efficient, robust, and affordable is also challenging. This does not mean there are no interesting candidates. Catalysts made from gold-copper bimetallic nanoparticles have shown promise, as too have hybrid systems that integrate bacterial enzymes with water-oxidizing catalysts.

The Future of Artificial Photosynthesis

Fossil fuels are in short supply and contribute to climate change. While alternative sources of energy exist, there are problems. Although we can convert the Sun's energy directly into usable electricity, we are unable to store this energy when the Sun is not shining. While it is possible to store and even convert this electrical energy, there will be loss of energy associated with that conversion. Replicating the process used by plants could solve some of these problems to create carbon-based fuels that will be renewable and carbon-neutral.

The major problem with the technology is that it is just a promise. While research is under way on making this a reality, scientists face many problems. While catalysts exist to oxidize water and reduce carbon dioxide, they are not cheap, abundant, or able to stand up to the conditions that they will be put under. The advantage, if the problem is solved, would mean that the technology could be incorporated into virtually any structure on Earth to provide a cheap, abundant, renewable, and carbon-neutral fuel source.

David Latchman is a freelance writer whose interest lies in writing mainly on physics-, chemistry-, and mathematics-related subjects to a lay audience. He holds a bachelor of science in physics with specializations in medical and environmental physics from the University of the West Indies.

Solar Energy for Remote and Rural Areas: Abhishek Rao

About 1.3 billion people globally—almost 20 percent of the world's population—have little or no access to electricity. These people live either in remote regions, far removed from the electricity distribution network called the "grid," or in rural areas of developing countries to which the electric grid infrastructure has not extended. They depend on diesel generators for their electricity needs, and on wood and kerosene as their primary sources of heating and lighting. Diesel, however, is a prohibitively expensive fuel, and is often difficult to distribute to remote and rural areas. Kerosene provides only dim, low-quality lighting; the burning of wood causes environmental pollution from carbon emissions, and both kerosene and wood cause chronic illnesses in humans from high levels of indoor air pollution. Improving access to cheap and clean energy increases rural income generation potential and enhances health and students' performance at school.

Extending the existing grid is an expensive task and is not feasible for remote, rural areas. In such situations, solar energy can power households and local businesses, by providing cheap and clean electricity generation and distribution at the individual house level or the village level. The simplest form of utilization of solar energy is the solar lantern, which costs under $25 and provides 2.5 hours of lighting or mobile-phone charging from charging the batteries for five hours in the Sun. Solar home systems (SHS) installed on the roof can power entire remote cabins or rural houses, including multiple lights, DC fans, and TVs. A basic SHS consists of solar photovoltaic (PV) panels, batteries, a charge controller that controls the battery charging from the panels, and possibly an inverter, depending on whether there are AC appliances in the house. This system costs $100, of which the PV panels account for 50% of the total cost, the batteries for 15%, and the remaining costs are for the wires and add-ons (Saxon Ghauri 2015, 14).

Un-electrified, rural households, on average, spend $2–$3 per month on kerosene and a lot of labor-hours gathering firewood. For $2 per month over two years, an off-grid consumer can get a solar lantern for basic lighting and mobile-charging purposes. A larger solar PV system that can serve 30 households together costs $1000, and can recover its cost by the solar energy it produces in one to two years. People in a group of houses can join together to pool their resources to purchase these off-grid PV systems, or they can draw individual or group loans from local banks or microfinance institutions at interest rates as low as 5 percent in some countries. Some other countries give out government subsidies for off-grid solar PV systems that make it easier for rural populations to own and install these systems and to start producing their own power.

While solar home systems power individual homes or a group of houses, decentralized renewable energy (DRE) systems are small solar power plants that generate close to the site of demand and power a number of households and businesses, and sometimes even entire villages or islands. SHS systems are generally 0.2–25 kW in size, and are called *pico-power* systems. In comparison, DRE systems or *mini-grids* have capacities greater than 25 kW. Besides powering multiple households or communities, DRE systems can provide power to commercial clients such as mobile telecommunication towers, rural ATMs, irrigation pumps, solar-powered refrigerators, and remotely located gasoline stations. In fact, commercial off-grid solar power applications such as these have three times the capacity of off-grid solar PV installed around the world than household consumers or village communities (Saxon Ghauri 2015, 3).

The primary disadvantage of using solar energy to power remote and rural areas is the intermittency of solar energy. SHS and DRE systems can produce power only when the Sun shines brightly, and not after dark or on cloudy or rainy days. This necessitates having batteries to store solar energy to be used at night, which increases maintenance costs if the batteries break down. Apart from battery backup, many off-grid solar PV systems are supplemented by standby diesel generating units, which can be switched on if the power from the PV panels is insufficient. Such hybrid off-grid power systems combining both solar PV and diesel generators are costlier but are more reliable. This is especially important for applications such as communication towers that must be powered all the time, and cannot depend solely on solar PV with battery backup.

While some people choose to live in off-grid houses or remote cabins in the woods, most remote and rural areas that are not connected to the national grid represent some of the poorest populations in the world. Without concerted efforts, the number of people with little or no access to electricity is not likely to drop. The onus to bring electricity to these people lies with national governments and public utilities that have the financial resources and technical capabilities for rural electrification programs. However, running remote, low-revenue-generating mini-grids in rural areas is often not a political priority for public electric utilities (Wiemann, et al. 2014, 28). In such circumstances, private-sector SHS and DRE system providers must develop business models that make off-grid solar PV systems with battery backup financially feasible. Growing awareness of the rural electrification problem in recent years, falling prices of PV panels and batteries, and greater access to finance are making off-grid solar energy economically attractive.

References

Saxon Ghauri, Clare, ed. 2015. "The Business Case for Off-Grid Energy in India." The Climate Group, in Partnership with Goldman Sachs and the Dutch Postcode Lottery. http://www.theclimategroup.org/_assets/files/ The-business-case-for-offgrid-energy-in-India.pdf. Accessed on April 25, 2015.

Wiemann, Marcus, Simon Rolland, and Guido Glania. 2014. "Hybrid Mini-Grids for Rural Electrification: Lessons Learned." Alliance for Rural Electrification, in Association with the United States Agency for International Development (USAID). http://ruralelec. org/fileadmin/DATA/Documents/06_Publications/ Hybrid_Mini-grids_for_Rural_Electrification_2014.pdf. Accessed on April 25, 2015.

Abhishek Rao is on a mission to help power the Earth with the Sun. He is a public advocate and published researcher in the renewable energy space. He holds a master's degree in solar energy engineering, business, and policy from Arizona State University.

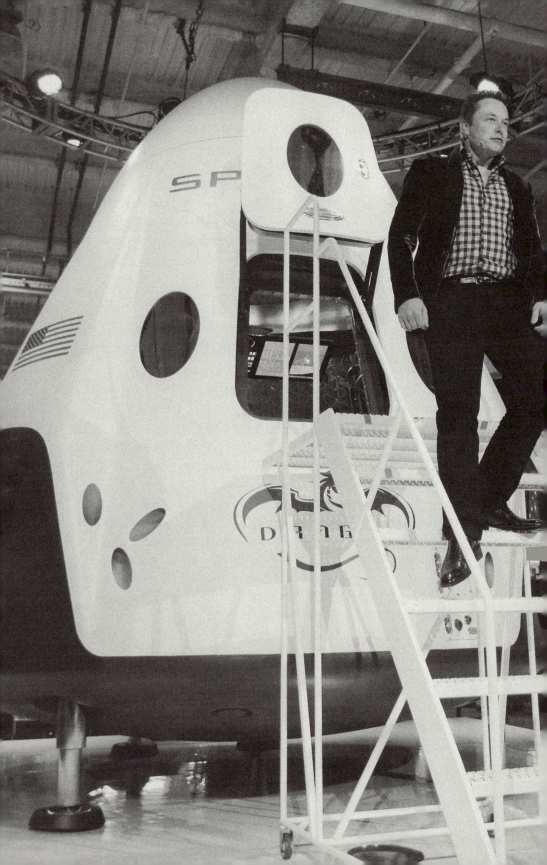

Introduction

One way to gain insight into the history and development of solar power, as well as to understand its place in current society and its promises for the future, is to learn about the men, women, and organizations who have been or continue to be actively engaged in the field. This chapter provides brief sketches of a number of individuals and organizations that have contributed to the development of solar power over the centuries; have been involved in the social, political, economic, environmental, and other aspects of solar power important in today's world; or are otherwise associated with this field of renewable energy.

William Grylls Adams (1836–1915)

Adams, working with his student Richard Evans Day, conducted a series of experiments on the photoconductive effects of the element selenium, confirming the fact that the element is capable of converting light energy directly into electrical energy. Adams and Day had read an earlier report by Willoughby Smith about his discovery of this phenomenon and decided to further pursue the effects that Smith had observed. The key experiment they conducted involved exposing a bar of selenium to a lit candle. They found that when the candle was brought

Elon Musk, CEO and CTO of SpaceX, walks down the steps while introducing the SpaceX Dragon V2 spaceship, which has electrical systems powered by solar energy. (AP Photo/Jae C. Hong)

to proximity with the selenium bar, the bar became conductive immediately. The significance of the experiment was that some questions had been raised about Smith's original work, asking whether it was heat or light that had produced the effects he observed. Since the selenium bar had no opportunity to warm up as a result of its exposure to the candle, Adams and Day concluded, it must have been the candle light (which travels at almost infinite speed) that had caused the phenomenon reported by Smith.

Williams Grylls Adams was born at Laneast, Cornwall, England, on February 16, 1836. His parents, Thomas and Tabitha (née Knill Grylls) Adams, were prosperous farmers in the area, and his older brother was John Couch Adams, the discoverer of the planet Neptune. Adams attended Birkenhead Park Grammar School in Cornwall before matriculating at St. John's College, Cambridge, from which he graduated in 1855. He then continued his studies at St. John's, eventually earning his BA, MA, and ScD in natural science, in 1856, 1859, and 1862, respectively. In 1863, Adams became a member of the Department of Natural Philosophy at King's College, London. Two years later, he succeeded James Clerk Maxwell as professor in that department, a post he held until his retirement in 1906.

During his lifetime, Adams's research covered a wide array of fields and topics, including the polarization of light, the flow of electric current through metal plates, the effects of magnetism on the flow of electric currents, the Earth's magnetic field, absorption of light in the atmosphere, and developments in lighthouse illumination systems. In addition to his scholarly research, Adams was also active in professional activities associated with his fields of research. In 1875, he was one of the founders of the Physical Society of London, of which he served as president from 1878 to 1880. In 1880, he was elected president of Section A (Mathematics and Physical Science) of the British Association for the Advancement of Science. And in 1884, he was chosen president of the Institution of Electrical Engineers. In 1872, Adams was honored by election to the Royal Society, where he later served on the administration council from 1882 to 1884 and again from 1896 to 1898.

Adams was a prolific writer, authoring a number of reports on his research. He may be best known, however, as editor of his brother John's scientific papers, a work of two volumes. He was also elected to be a fellow of the Geological Society and a fellow of the Cambridge Philosophical Society. After his retirement from King's College, Adams moved to Heathfield House, in Broadstone, Dorset, where he died on April 10, 1915.

Archimedes of Syracuse (ca. 287 BCE–ca. 212 BCE)

Some chroniclers of the history of solar energy point to Archimedes as the "first solar engineer," a accolade that must be taken with a very large grain of salt. The claim is based primarily on a single episode that supposedly illustrates the Greek scholar's understanding of the power of solar radiation to perform useful work. According to this story, Archimedes made use of solar radiation to destroy a fleet of Roman warships that had laid siege to the city of Syracuse, Archimedes's hometown, in 212 BCE. He is said to have laid out an array of reflecting mirrors in such a way that sunlight was reflected off the mirrors and onto the sails of ships in the Roman fleet in the harbor of Syracuse. The system worked so well, some historians suggest, that the sails caught fire and the fleet was destroyed.

Historians rely on reports of this event, some of which were written many years after it was supposed to have occurred. A number of problems are associated with the tale, not least of which is the fact that the siege was eventually successful, Syracuse fell to the Romans, and Archimedes himself was killed by a Roman soldier in the streets of the city. Nonetheless, countless experiments have been designed and conducted by scholars looking for proof that the scientific principle involved in this tale is, at the very least, valid. Generally speaking, the consensus appears to be that it was not, and that the geometry and mechanics of a solar array sufficient to produce such widespread destruction was not available until at least a century after Archimedes's death, with the discovery of parabolic mirrors by the Greek scholar Diocles in the second century BCE.

Scholars know almost nothing for certain about Archimedes's life, other than that his father, Phidias, was an astronomer. Everything else claimed for his life is based on secondhand accounts, some of which were repeated a number of times over the years, only adding to the uncertainty of the information they presumably provide. One of the most familiar stories about Archimedes tells of his trip to Egypt, where he invented a device for lifting water out of ditches for use in irrigation system, a device now known as an Archimedes screw. Perhaps the best-known story about Archimedes relates to his supposed discovery of the principle of buoyancy. According to that story, King Hiero II had asked that a new crown be made for him, but when the crown arrived, he suspected that it was not made of pure gold, as he had directed. Hiero asked Archimedes to find a way of determining the composition of the crown without, of course, breaking it apart and analyzing the material of which it is made. The story is told that Archimedes discovered the solution to this problem as he was sitting in his bath one day, noting by chance that he displaced a certain amount of water as he entered the bath. He concluded that a similar method could be used to determine the density (and hence the composition) of Heiro's crown. As soon as he realized he had reached his answer to Heiro's challenge, Archimedes jumped from his bath and ran through the streets of the town naked shouting "Eureka!" (which means "I've got it!").

Another of Archimedes's famous discovery is thought to be the law of the lever. A lever is one of the six simple machines on which virtually all other mechanical devices are based. Archimedes found in his research on the lever the fundamental principle that the mass of an object multiplied by its distance from the fulcrum of the lever is equal for both arms of the lever. The principle means that the lever can be used to lift heavy weights with relatively modest effort provided they are placed at a specific distance from the fulcrum. In describing his law to contemporaries, he is said to have claimed that he could perform prodigious feats making use of his new law. "Give me a place to stand," he said, "and I can move the world." (One can only imagine the length of the lever arm that would be needed for such an act.)

Archimedes was particularly proud of his achievements in mathematics, and scholars believe it is the only field in which he actually published his work (dealing with physical reality was too "mundane" for a true philosopher, in his eyes). He turned his attention, for example, to calculating the value of pi, one of the problems of special interest to early mathematicians. He was never able to provide a precise value for the number but did conclude that it must lie somewhere between 223/71 and 22/7 (3.1408 and 3.1428), a remarkably good value for that period in history. In one of his most ambitious efforts, Archimedes also attempted to calculate the number of grains of sand needed to fill the universe, a number that would today be expressed as 8×10^{63}.

Archimedes is thought to have died in 212 BCE during an attack on the city of Syracuse by Roman troops that he was attempting to prevent with his attack with solar energy on the enemy fleet. The story is told that he was busy making calculations in the sand when he was accosted by a Roman soldier. Supremely oblivious to the danger in which he stood, Archimedes simply warned the soldier, "Don't disturb my circles." Somewhat less concerned with the value of Archimedes to human history, the soldier is said to have run him through with his sword, thus bringing to a close one of the great lives in human history.

American Council on Renewable Energy

The American Council on Renewable Energy (ACORE) is a trade association consisting of more than 500 companies, industry associations, utilities, end users, financial and educational institutions, professional service firms, nonprofit organizations, and government agencies who are involved in one way or another in the production, distribution, financing, use, and other aspects of renewable energy. Examples of current members include Acciona Energy North America; American Clean Energy; Bostonia Partners; Carlyle Capital Markets; City of Palo Alto Utilities; Duke Energy; Everpower Wind Holdings; Franklin and Marshall College; Green Strategies, Incl.; Ionex Energy Storage Systems; NRG Energy; POET Biorefining; Professional Engineers in California Government; Royal Danish Embassy (Washington, DC); Trina Solar Energy Corporation; and Wells Fargo—Environmental Finance. The organization's stated mission is "to mov[e] renewable energy into the mainstream of America's economy, ensuring the success of the renewable energy industry while helping to build a sustainable and independent energy future for the nation."

ACORE was founded in 2001 for the purpose of bringing together groups of all kinds interested in moving renewable energy into the mainstream of America's economy. The organization's current activities are centered in six major programs:

- The ACORE Leadership Council provides a mechanism by which experts in various fields are able to express their views on essential issues in the field of renewable energy. The council has produced a number of papers on such issues and provides speakers on a variety of related topics.

- The Biomass Coordinating Council works to increase the role of biomass energy in the nation's energy equation, thus reducing its dependence on foreign fossil fuels. The council sponsors working groups, webinars, regional roundtables, and other educational events for the dissemination and discussion of topics related to biomass energy.

- The REFIN Directory is a tool for connecting groups with financial, personnel, and other resources with companies in need of such resources. As an example, REFIN made possible the financing of three wind energy projects by Citibank.

- ACORE Regional Outreach Program is based on the fact that energy projects are seldom based on the geography of state boundaries but generally extended over wider areas. The outreach program attempts to bring together information about energy issues across larger regions (such as the West, Midwest, and Southeast) to facilitate the development and use of renewable energy resources.

- U.S. Partnership for Renewable Energy Financing (US PREF) is a program of experts from financial institutions, technology companies, project development companies, law firms, and other organizations to discuss policies for the advancement of renewable energy development in the United States.

- U.S. China Program is a joint project between ACORE and Chinese groups, such as the Chinese Renewable Energy Industries Association (CREIA), for the fostering of cooperation between agencies and companies in the two countries with regard to renewable energy projects.

The primary mechanism through which ACORE works toward its goals are annual conferences that focus on three topics: marketing, finance, and policy. The RETECH marketing conferences bring together business leaders, investors, renewable company representatives, state and federal government officials, college and university educators, and exhibitors to discuss and learn about advances in wind, solar, geothermal, biomass, and other renewable energy technologies and supporting technologies. The 2010 RETECH conference included more than 3,000 attendees. The finance conferences provide executives of renewable energy companies with information about the ways to obtain financing for their projects and to make sure that those projects reach a successful conclusion. The policy conferences deal with all aspects of the production, storage, transmission, and consumption of all forms of renewable energy. The 2011 conference in Washington, D.C., for example, dealt with topics such as the effect of the current economic slowdown on the renewable energy industry, international growth in the field with related effects on competition by American industry, increasing demands by the Department of Defense for renewable energy in its activities, and possible effects of the expiration of biofuels incentives at the end of the year. The 2011 U.S.-China workshop on renewable energy was held in Beijing on November 1. Topics included ways in which American companies can learn more about foreign investment in renewable energy based on the very successful Chinese experience, the very successful Chinese solar market and ways in which American companies may be able to partner with Chinese companies in this field, and the broader picture of renewable energy development in China and its implications for the U.S. industry.

ACORE annually publishes a variety of articles, reports, white papers, and other publications dealing with all aspects of renewable energy. Some examples include the article "Renewable Energy Trends 2010," for the magazine *Infrastructure Solutions*; the biannual report "Renewable Energy in America: Markets, Economic Development and Policy in the 50 States;" the slide collection "Renewable Energy Landscape: Market & Technology Overview;" the report "Renewable Energy Landscape: Market & Technology Overview;" and the conference report "RETECH 2009 Summary Report." Of particular interest and value to the general reader is ACORE's compilation of statistics on the renewable energy industry, which is available as a slide show online. The slide show is available free to members of the organization and for sale to nonmembers. A sample version of the product is also available online at http://www.acore.org/publications/renewable-energy-landscape.

ACORE's website also provides a somewhat abbreviated various types of renewable energy, such as biofuel, biomass, geothermal, hydroelectric power, hydrogen energy, ocean and tidal energy, solar energy, waste to energy, and wind power. The website also has a useful and extensive section on careers in renewable energy.

American Solar Energy Society

The American Solar Energy Society (ASES) was founded in 1954 to work toward a greater production and use of solar energy in the United States and throughout the world. The organization was the outgrowth of an earlier association known as the Association for Applied Solar Energy (AFASE) and later evolved into a larger and most comprehensive group known as the International Solar Energy Society (ISES). For further information about the international group, see the heading in this chapter under that name.

Today, ASES is organized in two ways, first as a group of state and regional chapters and as a group of divisions devoted to specific aspects of solar energy. In general, most states have at least one chapter as part of ASES overall, while some states contain more than one chapter, and other states are part of regional chapters. For example, New York State has its own state chapter, New York Solar Energy Society, Inc., as well as a regional chapter, Northeast Sustainable Energy Association. North Carolina has its own state chapter, the North Carolina Sustainable Energy Association, along with two student chapters at Appalachian State University and North Carolina State University. The technical divisions within ASES focus on solar electric, solar buildings, solar thermal, sustainable transportation, concentrating solar power, small wind, resource assessment, clean energy and water, and sustainability. Typical of the activities conducted by each division are the publishing of newsletters in the special area of interest, organizing webinars, conducting an annual meeting in conjunction with the ASES annual conference, providing technical expertise in its own area of interest, and keeping members informed of news and developments in that area.

In addition to individual memberships, corporate memberships in ASES are also available. Some of the 141 current corporate members of the organization are Alfred (New York) State College, Aztec Renewable Energy (Dallas), Brightergy

(Kansas City, Missouri), Cemaer (Mexico), EcoSolargy (Irvine, California), Everglades University (Florida), Nevada State Office of Energy, Roof Power Solar (Rich Hill, Missouri), Southern California Edison, and Zager Plumbing & Solar (Deerfield Beach, Florida).

From an educational and promotional standpoint, the major ASES event of the year is its National Solar Tour. The tour provides an opportunity for individuals to visit homes to see the variety of ways in which solar technology can be used in individual residences. The 20th annual National Solar Tour was held in 2015, with more than 150,000 participants as guests at more than 5,000 private homes. More information about the tour with an interactive map of participating residences is available at the ASES website at http://www.ases.org/solar-tour/.

The other major ASES event of the year is its annual conference, Solar XXX, where XXX represents the year of the conference. At Solar 2014, held in San Francisco, some of the topics discussed were advances in passive solar design, emerging architecture, concepts in PV, solar power forecast applications, passive and net zero energy homes, emerging transportation, and women in renewables.

Alexandre-Edmond Becquerel (1820–1891)

Becquerel is acknowledged as the discoverer of the photovoltaic effect, the process by which a beam of light impinges on a metallic material, causing the release of electrons in a flow of electric current. The principle is of unparalleled importance in the production of solar power, of course, and is the primary mechanism by which photovoltaic (PV) cells operated. Because of this work, Becquerel is sometimes known as the father of the photovoltaic cell.

Becquerel came from one of the most prestigious families of French scientists in the nineteenth and twentieth centuries. His father was Antoine César Becquerel, who had made a number of important discoveries in the fields of electricity and luminescence, and who had professor of physics at the Muséum National d'Histoire Naturelle from 1837 to 1878. Alexandre-Edmond's son, Henri Becquerel, followed in the tradition of his father and grandfather, working primarily in the field of physics, where he is best known for his discovery of radioactivity, for which he earned the 1903 Nobel Prize in Physics.

Alexandre-Edmond Becquerel was born in Paris on March 24, 1840. He received his secondary education at the Lycée Bourbon (now the Lycée Condorcet). At the age of 18, he passed qualifying examinations for and was admitted to both the École Normale Supérieure and the École Polytechnique, two of the most prestigious institutions of higher learning in France. He chose instead, however, to accept an offer to work as an assistant to his father at the Muséum National d'Histoire Naturelle, a post he held for many years, eventually succeeding his father as professor of physics at the Museum in 1878. Upon Alexandre-Edmond's own death in 1891, he was succeeded in the chair of physics by his own son, Henri, who held the post until his own death in 1908. The heritage continued one more time, then, when Henri's son, Jean-Antoine-Edmond Berzelius, succeeded to the chair of physics at the museum in 1909. (That line of succession ended when Jean-Antoine-Edmond died in 1953 without an heir.)

In 1849, Edmond Becquerel accepted an appointment as professor at the newly created National Agronomic Institute at Versailles. The institute opened in 1850 with great hopes that it would become one of the leading agricultural schools in the nation. However, it lasted less than two years, largely because its costs were far greater than had been anticipated. Becquerel did not remain long with an academic position, however, as he was then appointed in 1853 to the chair of physics at Conservatoire des Arts et Métiers in Paris, where he remained until he succeeded his father at the Muséum National d'Histoire Naturelle five years later.

Becquerel was especially interested in the nature of light and its effects on chemical processes. Among his most active areas of research was the subject of phosphorescence, the process by which a substance gives off light without undergoing combustion. As a part of this line of research, he invented a device known as a phosphoroscope that can be used to measure the amount of time that passes between the exposure of a substance to a beam of light and the time at which phosphorescence is first observable.

Becquerel's most famous publication was a two-volume work on the nature of light, *La lumière, ses causes et ses effets* (*Light, Its Causes and Effects*), published in 1867 and 1868. He died in Paris on May 11, 1891, at the age of 71. In 1989, the European Commission created the European Becquerel Prize for Outstanding Merits in Photovoltaics in his honor. The award is given annually "to honor scientific, technical or managerial merit in the development of photovoltaic solar energy, attained over a long period of continuous achievements, or very exceptionally, for some extraordinary invention or discovery."

Jimmy Carter (1924–)

Carter was arguably the most influential advocate for solar energy in the United States during the first half of the 1970s. As president of the United States, Carter laid out a plan for encouraging and supporting the growth of solar power at a time when the nation and the world was faced with a sudden and dramatic loss of access to traditional fossil fuels as a result of the Arab oil embargo of 1973. Only months after he was elected president in 1977, Carter laid out his plans for promoting the growth of renewable energy, including solar power, in a televised speech to the American public. Two years later, he made a major policy speech about solar energy to the U.S. Congress in which he set out the argument for expanding the use of solar power. "Energy from the sun is clean and safe," he said. "It will not pollute the air we breathe or the water we drink. It does not run the risk of an accident which may threaten the health or life of our citizens. There are no toxic wastes to cause disposal problems." He announced that he planned to ask for $1 billion for research and development of solar power in the United States in FY 1980. In addition, he said that he was creating a Solar Bank to make it easier to obtain financing for solar projects. The bank was to be funded at an initial level of $100 million and was to operate out of the Department of Housing and Urban Development. He also outlined changes in federal tax policy providing tax credits and other incentives for the purchase of equipment used in the development of solar power. Finally, in a later action with more symbolic than practical significance, Carter ordered the installation of 32 solar panels on the roof of the White House on the day of his historic speech to Congress.

James Earl Carter, Jr. (generally known as "Jimmy") was born in Plains, Georgia, on October 1, 1924. His mother was Bessie Lillian Gordy Carter, a registered nurse, and his father was James Earl Carter, Sr., a farmer and businessman who ran a general store. In 1928, the Carters moved from Plains to the nearby town of Archery, a community consisting largely of African Americans. Although James Carter, Sr., still believed in segregation of the races, he allowed his son to befriend his neighbors regardless of their skin color. Jimmy became an entrepreneur at a young age when he began growing peanuts on an acre of land that his father had given him. He conducted his business while attending Plains High School, from which he graduated in 1941.

After graduation, Carter matriculated at Georgia Southwestern College in nearby Americus. He regarded his time at Southwestern primarily as preparation for his real ambition, obtaining an appointment to the U.S. Naval Academy in Annapolis, Maryland. Thus, he focused on the courses at Southwestern that would allow him to qualify for the Naval Academy and not necessarily for graduation from Southwestern. Finally, in 1943, Carter obtained his appointment to the

Naval Academy, from which he graduated in 1946. After receiving his commission, Carter was assigned to the battleship USS *Wyoming*, in Norfolk, Virginia. A month after receiving his assignment, he was married to Rosalynn Smith, with whom he was eventually to have four children, John, James III (Chip), Donnell, and Amy.

In 1948, Carter was accepted at the submarine officer training school at New London, Connecticut, setting the path he had decided to follow in the navy. After assignments in Honolulu and San Diego, Carter returned to New London in 1952, where he had been accepted as a member of Admiral Hyman Rickover's new nuclear submarine program. At the completion of the program, Carter was assigned to the USS *Seawolf*, one of the first U.S. nuclear submarines.

At this point in his life, Carter seemed set on a career in the U.S. Navy, with the possibility of significant advances up the career ladder. That future was interrupted, however, with the death of his father on July 23, 1953. After agonizing over his choices, Carter finally decided to resign his commission in the navy in order to return to Georgia and take up the running of his father's business. He left the navy with an honorable discharge on October 9, 1953, and returned to Plains.

For the next decade, Carter struggled with the peanut business in Plains, dealing with a variety of economic and environmental issues that nearly brought the business to bankruptcy. As he struggled with the business, Carter also became involved with dramatic political changes occurring around him as the result of the U.S. Supreme Court's 1962 *Baker v. Carr* case, in which the Court affirmed the now-famous "one-man, one-vote" policy for all elections in the United States. That decision led to the opening of elections in Georgia that had previously been controlled by a handful of influential individuals. Carter decided to run for the state senate at the last moment and, when votes had been counted, appeared to have lost the race. A recount found that questionable voting had occurred and that instead of losing by 139 votes, he had actually won by 831 votes. Carter took his seat in the Georgia senate on January 14, 1963.

Carter was reelected to the state senate in 1964 but began to think of higher offices even before his second term ended. In 1966, he announced that he had decided to run for the U.S. House of Representatives, although he soon changed his mind and decided to run for governor of Georgia instead. Carter lost that election and returned to the peanut farm in Plains. But he had announced to his supporters that he would be back and, in fact, he ran for governor again in 1970, this time successfully. A month before completing his first term as governor (Georgia governors are allowed to serve only one term), Carter announced that he was running for president of the United States in the 1976 election. At the time, he was virtually unknown nationally and was thought to have no chance of success. He pursued an aggressive campaign, however, and was elected 39th president of the United States on November 2, 1976.

Carter served only one term as president, being defeated in a landslide by Ronald Reagan in 1980. He then returned to Plains but began a second career that has involved a wide array of political and humanitarian activities, including visits to almost every part of the world to encourage fair elections and humane treatment of all groups of individuals. In 2002 Carter received the Nobel Peace Prize in recognition of his efforts "to find peaceful solutions to international conflicts, to advance democracy and human rights, and to promote economic and social development."

Jan Czochralski (1885–1953)

Czochralski is widely said to be the most cited of all Polish scientists and one of the three best-known scientists from that nation (with Marie Sklodowska-Curie and Nicolaus Copernicus being the other two). He is best known for having discovered a method for growing very pure metallic single crystals, a technology that is essential to the manufacture of solar photovoltaic cells today. The method is named after him, the Czochralski method. The method is very simple in concept but somewhat complicated in actual process. It begins by attaching a seed crystal of the material to be made to a long wire, which is then dipped into a molten sample of that material. The wire is then withdrawn very slowly from the melt at such a pace as to allow the molten material to cool and condense on the solid seed crystal. When carried out properly, the product of this procedure is a large mass of the original molten material in the form of a single crystal with a (usually) very pure composition.

Jan Czochralski was born in Exin, in what was then the Prussian province of Pomerania, and is now Kcynia, Poland, on October 23, 1885. He was the eighth child born to Franciszek Czochralski and Marta Suchomski Czochralski. At his parents' suggestion, Czochralski completed a seminar for the preparation of teachers in Kcynia. He completed the course successfully but with such poor grades that he did not receive his matriculation certificate. Without that certificate, he was unable to continue his education, so decided to move to the nearby city of Krotoszyn, where he took a job as an assistant in a drugstore and continued to study chemistry on his own.

In 1904, Czochralski decided to pursue his fortunes in Germany and moved to Berlin, where he once again took a job in a pharmacy. Shortly thereafter, he took a position with Kunheim and Co. in Niederschöneweide, now a part of Berlin, a chemical manufacturing company. In August 1907, he then moved to another industrial corporation, the Allgemeine Elektrizitäts-Gesellschaft (AEG), where he remained until 1917. He was eventually promoted to head of the steel and iron research laboratory at the Kabelwerk Oberspree plant of AEG. During the time that he was developing his skills in working with a variety of metals at these industrial firms, Czochralski was also taking classes in chemistry at the Charlottenburg Polytechnic in Berlin. It was during his tenure at AEG that Czochralski accidentally discovered his method for making large, single crystals of metals. It is said that he absentmindedly dipped his pen into a container of molten tin, rather than into an inkwell, and noticed as he withdrew the pen that the molten tin solidified to form a large, nearly perfect metallic crystal. The accidental discovery inspired Czochralski to explore under controlled circumstances in more detail the process that he had observed, eventually resulting in his classic paper about the discovery, "Ein neues Verfahren zur Messung der Kristallisationsgeschwindigkeit der Metalle" ("A new method for the measurement of the crystallization rate of metals), published in the *Zeitschrift für Physikalische Chemie* (the *Journal of Physical Chemistry*) in 1918.

Czochralski's success with his new process prompted him to contact the Metallbank und Metallurgische Gesellschaft AG company in Frankfurt am Main about establishing a new metallurgical laboratory committed to both research and development of new and useful metallurgical technologies. He became head of the new facility, a post he held until 1928. He then left Germany at the request of the Polish government to return home and assume the post of professor of metallurgy and metal research in the chemistry department at the Warsaw University of Technology. He also served as head of the newly created Metallurgy Department of the Industrial Chemistry Research Institute in Warsaw.

At the conclusion of World War II, Czochralski was accused by the new Communist government of collaborating with German authorities during the war. Although a government investigator eventually concluded that there was no basis for this charge, the Technical University refused to rehire Czochralski. He decided to return to his hometown of Kcynia in 1945, where he established a small cosmetics and household chemicals business. He died in Poznan on April 22, 1953. On June 29, 2011, the Technical University corrected its long-held view about the role Czochralski played in World War II, adopting a resolution restoring his good name at the institution.

Horace-Bénédict de Saussure (1740–1799)

De Saussure is best known in the history of solar energy for his studies of solar "hot boxes," devices consisting of nested wooden boxes with black interiors covered by sheet glass. When placed in the sun, such boxes could receive substantial temperatures, sufficient to boil water and maintain elevated temperatures. Although such devices may now seem to be very simple in design, they represented an important step forward in the attempt to capture and concentrate solar energy. De Saussure's inventions inspired later generations of inventors to improve on his own devices, eventually leading to large, powerful solar engines with the capability of performing a variety of large-scale work projects. De Saussure is also credited with inventing a device for measuring the amount of heat produced by the sun in a particular location, a device called a *heliothermometer*. The heliothermometer had essentially the same structure as did one of his hot boxes, although de Saussure used the device for other purposes.

Horace-Bénédict de Saussure was born in the Swiss region of Conches, adjacent to Lake Geneva, on February 17, 1740. Although his home was near the city of Geneva, he spent nearly all of his life to the age of 25 at his country estate near the lake. He later commented in one of his works that such an upbringing "seem[ed] made to inspire a taste for Natural History. Nature presents herself in her most brilliant aspect." It was hardly surprising, then, that he chose to spend his life in a study of nature, partly as a physicist, and partly as an alpinist, a vocation of which he is said to have been the founder.

De Saussure was sent to the local primary school at the age of six, where he achieved some early success and gained a prize in a reading competition at the school. At the age of 14, he then matriculated at the local academy, the equivalent of a modern-day university, where his experience was not so salubrious. In fact, his biographer has written that de Saussure "retained no favourable recollections" of his years at the academy and was sufficiently unsatisfactory to convince him not to send his own children to the school. He apparently took advantage of every moment available away from classes to explore the mountainous region in which his home was located. It was at this early age already that he was beginning to appreciate the wonders of mountain exploration that lies at the heart of alpinism.

In fact, exploration of the flora, fauna, and geology of the Alpine region was the dominating theme of the rest of de Saussure's life. His biography is to a large extent a record of the mountains he climbed and the research he conducted during these expeditions. He wrote in particular about the rocks and minerals he found, the geological conformations in which they occurred, their chemical structure, the fossils they contained, the weather conditions he encountered on his travels, and the information these data provided about the origin of the Earth in general and the Alpine region in particular. His contributions to the sciences are recognized today in a mineral named for him, saussurite, and a genus of alpine plants, *Saussurea*. In addition, the Alpine Botanical Garden Saussurea located at Pavillona du Mont Fréty is named in his honor.

Although he has a confirmed place in the history of solar energy, de Saussure's "hot boxes" really had a very limited purpose that had virtually nothing to do with the use of solar energy for practical purposes. Thus it is probably his impact on later inventors for which any honor in solar history depends.

For much of his life, de Saussure was forced to relegate his true love for alpinism to a "real" career in academia. He was appointed professor of metaphysics at the University of Geneva in 1762 at the young age of 22, a post he held for a quarter of a century, retiring in 1786. The position was not an easy one as individuals who held that appointment were expected to lecture on natural science in French and the next year on metaphysics in Latin. With this type of assignment, one can easily imagine de Saussure's disappointment in having less time than he might otherwise have enjoyed in his beloved Alps.

After retiring from the university, de Saussure continued his exploration of the alpine region until nearly the last year of his life. He died in Geneva on January 22, 1799.

Peter Glaser (1923–2014)

Glaser was a Czechoslovakian-born American aerospace engineer who proposed the concept of space-based solar systems in the late 1960s. He laid out the principles behind his plan for collecting solar power from satellites orbiting Earth and transmitting it to Earth's surface in an article published in the journal *Science* in November 1968 ("Power from the Sun: Its Future." *Science*. 162(3856): 857–861). In that article, he presented the argument that human dependence on fossil fuels was inevitably destined to end at some point in the future and that, no matter how soon or how late that point was to appear, nations would have to be prepared to find alternative sources of energy other than coal, oil, and natural gas. He suggested that solar radiation would be an excellent candidate for that alternative and renewable source of energy and that it would behoove researchers to begin developing the technology for harvesting sunlight with satellites. Glaser's proposal was sufficiently convincing that the U.S. government eventually committed about $20 million to studying satellite solar power systems, although it eventually concluded that the concept was not economically feasible. Glaser did receive a patent on the fundamental technology involved in 1973.

Peter Edward Glaser was born in Zatec, Czechoslovakia, now part of the Czech Republic, on September 5, 1923, to Hugo and Helen (Weiss) Glaser. During World War II, his family fled to England, where Glaser joined the Free Czechoslovak Army as a tank commander. During and after the war he attended the Leeds College of Technology in England and Charles University in Prague, from which he received diplomas in 1943 and 1947 respectively. He then continued his studies at Columbia University in New York City, where he received his master of science and doctor of philosophy degrees in 1951 and 1995, respectively, all in the field of mechanical engineering. While working for his graduate degrees, Glaser was head of the design department at Werner Management Company in New York City from 1948 to 1953. He became a citizen of the United States in 1954.

After receiving his PhD, Glaser accepted an offer to work at Arthur D. Little, a management and consulting firm then based in Boston, Massachusetts. He remained with them for his entire professional career, finally retiring in 1999. During his time at the company, Glaser was involved in a wide variety of projects, including high temperature research, remote sensing systems, solar and arc imaging furnaces, solar heating and cooling, photovoltaic cells, rural electrification systems, renewable energy resources, lunar space programs, commercial space power, climate change issues, space station technology, and space travel equipment. He was intimately involved in the planning for the *Apollo* space missions and other space exploration missions, serving on a number of committees, such as the National Aeronautics Space and Administration (NASA) committees on Task Force on Space Goals and Lunar Energy Enterprise Case Study Task Force, and the Space Power Committee of the International Astronautical Federation. He served as project manager for the *Apollo 11* Laser Ranging Retroreflector Array and for the Lunar Heat Flow Probes and the Initial Blood Storage Experiment. Glaser was also president of the International Solar Energy Society from 1968 to 1969 and editor in chief of the journal *Solar Energy* from 1971 to 1984.

After the U.S. government abandoned its interest in satellite solar power, Glaser founded a new nonprofit organization for the study of the technology, the SUNSAT Energy Council, which became affiliated with the United Nations Economic and Social Council. Glaser served as both president (1974–1994) and CEO (1994–2005) of SUNSAT. Although that organization has now become largely defunct, other nations have adopted the idea of satellite solar power, perhaps none so aggressively as the Japanese. In 2015, Japanese researchers announced the successful implementation of a wireless system for transmitting solar power from satellites to stations on Earth, the first concrete realization of Glaser's nearly 40-year-old dream.

Glaser was active in a number of space-related organizations. He was a fellow of the American Association for the Advancement of Science (AAAS) and the American Institute for Aeronautics and Astronautics as well as a member of the American Society of Mechanical Engineering, the American Astronautical Society (AAS), the National Space Society (NSS), and the International Astronautical Federation (IAF). He was the recipient of the Carl F. Kayan medal from Columbia University in 1974 and the Farrington Daniels Award of the International Solar Energy Society in 1983. In 1993, the NSS and the IAF established the Peter Glaser Plenary Lecture to be given at the organizations' joint annual congress. Glaser was also a regent of the United Societies in Space (USS). He was inducted into the Space Technology Hall of Fame of the U.S. Space Foundation (USSF) in 1996.

Glaser died at his home in Lexington, Massachusetts, on May 29, 2014, at the age of 90.

International Energy Agency

The International Energy Agency (IEA) was founded in 1974 in response to the world crisis in energy supplies resulting from the oil embargo announced by the Organization of Arab Petroleum Exporting Countries (OAPEC). The embargo had been established as a way by which Arab states could object to support for the state of Israel by the United States and other Western nations. Confronted with the sudden and drastic loss of its most important energy source, petroleum, a number of nations banded together to form the International Energy Agency. IEA's initial goals were to develop more sophisticated and accurate statistics on known and estimated petroleum reserves and oil production and consumption and to investigate ways of ameliorating the effects of the dramatic loss in petroleum supplies for developing nations. Since 1974, IEA's mission has expanded somewhat and now includes four major themes.

The first of these themes is Energy Security, which involves the promotion of diversity and flexibility of energy supplies in order to avoid the worst consequences of fuel crises like that of the 1974 oil embargo. The second theme is Environmental Protection, which recognizes the deleterious effects of extensive fossil fuel use. A recent feature of this theme is a greater emphasis on the possible global climate effects resulting from the combustion of coal, oil, and natural gas. The third theme is Economic Growth, which emphasizes the importance of developing energy policies that ensure the continued economic development of all nations, whether developed or developing. The fourth theme is Engagement Worldwide, which is a specific acknowledgment that the energy and environmental policies facing nations of the world transcend national borders and can only be solved by international cooperation.

Only member states of the Organisation for Economic Co-operation and Development (OECD) are permitted to belong to the IEA. Current members are Australia, Austria, Belgium, Canada, Czech Republic, Denmark, Estonia, Finland, France, Germany, Greece, Hungary, Ireland, Italy, Japan, Luxembourg, the Netherlands, New Zealand, Norway, Poland, Portugal, Slovakia, South Korea, Spain, Sweden, Switzerland, Turkey, the United Kingdom, and the United States. Although prohibited from inviting non-OECD nations to join the agency, the IEA has a well-developed program for working with such nations through its Directorate of Global Energy Dialogue (GED). Established in 1993, the GED attempts to better understand the status of energy systems in non-OECD nations and to share with those nations some of the knowledge and expertise developed through IEA programs. The agency currently has bilateral and regional agreements with a number of non-OECD nations and regions to achieve these goals, including Brazil, Caspian and Central Asia, Central and Eastern

Europe, China, India, Mexico, the Middle East, Russia, Southeast Asia, Ukraine, and Venezuela.

The primary decision-making body of the IEA is the Governing Board, which consists of the energy ministers of all member states, or their designated representatives. Decisions made by the Governing Board are carried out by the Secretariat, which consists of researchers and other scholars. The Secretariat's work involves the collection of data on energy supplies, production, and consumption; sponsorship of conferences and other meetings on energy issues; assessment of energy conditions in member states; development of projections for global energy futures; and proposals for national energy policies in member states.

The work of the Secretariat is organized around a number of specific topics, for which a variety of reports and publications are generally available. The current topics of interest to the IEA include coal, carbon dioxide capture and storage, cleaner fossil fuels, climate change, electricity, energy efficiency, energy indicators, energy policy, energy predictions, energy statistics, fusion power, greenhouse gases, natural gas, oil, renewable energy, sustainable development, and technology. For the general public and experts in the field, IEA's most important contributions may well be their regular reports on a variety of important energy-related issues. In the area of statistics, the agency publishes a host of reports on a monthly, quarterly, and annual basis. Perhaps best known of these is its annual *Key World Energy Statistics*, published regularly for the last 10 years. Other statistical publications deal with energy prices and taxes, energy statistics for OECD and non-OECD states, oil information, natural gas information, renewables information, and carbon dioxide emissions from fuel combustion. The agency also publishes a quarterly publication, *Oil, Gas, Coal and Electricity*.

Other fields for which publications are available are Global Energy Dialogue, which includes publications on energy issues in non-OECD nations; Energy and Environment, which deals with topics such as the environmental and climate effects of fossil fuel combustion; Renewable Energy, which considers technical, social, economic, and political aspects of the development of renewable energy sources; Energy Efficiency, which discusses ways in which conservation can extend the lifetime of existing fossil fuel resources; Energy Technology Network, which reports on agreements made among member states to improve their energy production and consumption patterns; Policy Analysis and Cooperation, which reviews energy policies, practices, and projections for specific member states; and Energy Technology, which reviews proposed and in-place changes in technology to make more efficient use of existing and proposed energy sources. An especially interesting feature of the agency's website is its "Fast Facts" section, which lists tidbits of information about many aspects of energy. Each item is cited from an IEA publication, to which a link is provided in each case.

In addition to its print publications, the IEA sponsors a number of conferences and other meetings, provides speakers for a variety of professional and community events, and testifies before governmental agencies in member nations. In 2011, for example, the agency sponsored, cosponsored, or participated in meetings on energy outlooks, in Saudi Arabia; energy storage, in Paris; renewable energy, in New Delhi; electric vehicles, in Shanghai; a clean energy symposium, in Astana, Kazakhstan; energy efficiency in buildings, in Brussels; and low-income weatherization projects, in Dublin.

International Solar Energy Society

The International Solar Energy Society (ISES) claims to be "the longest standing solar organization in the world." It was founded in 1954 in Phoenix, Arizona, as a nonprofit organization called the Association for Applied Solar Energy (AFASE). According to a 1985 article in the journal *Technology and Culture*, the organization was created to "generate public interest in solar energy development, expand research, and encourage the commercial application of solar power to industrial and residential needs." In 1955, the association held two highly successful conferences in Phoenix that were attended by more than a thousand scientists, engineers, and government officials from 36 countries. A year later, AFASE released its first publication about solar energy, a journal titled *The Sun at Work*, a publication that remained in print until 1969. In 1957, the association also published its first scholarly journal, *The Journal of Solar Energy, Science and Engineering*.

By 1963, members of the AFASE had begun to rethink the scope of their mission. They had decided that the goals they wished to achieve would be better met by creating an international organization, resulting in a renaming of the association as the Solar Energy Society. In recognition of their new international approach, the group also requested and received recognition by the United Nations Economic and Social Council. A year later, the group also changed the name of its journal to *Solar Energy: The Journal of Solar Energy Science and Technology*. By 1970, the association's international character was further recognized by moving its headquarters from Phoenix to Melbourne, Australia, and, a year later, by changing its name to the International Solar Energy Society.

Today, the ISES consists of more than 50 national and regional sections in over 100 countries in every part of the world. Some examples of these sections are the Egyptian Association for Energy and Environment and Sustainable Energy Society of South Africa, in Africa; Australian Photovoltaic Institute and Sustainable Energy Industry—Pacific Islands, in Oceania; Asociación Argentina de Energías Renovables y Ambiente and Associacao Brasileira de Energia Solar, in South America; ISES Bulgaria and Danish Solar Energy Society, in Europe; Chinese Renewable Energy Society and Solar Energy Society of India, in Asia; and Solar Energy Society of Canada, Inc. and American Solar Energy Society, in North America. In addition, a section called Young ISES is available to students, graduate students, and young professionals interested in the field of solar energy. The ISES has also developed working relationships with a number of corporate and nonprofit organizations interested in solar energy, such as the American Council on Renewable Energy and the International Renewable Energy Alliance.

ISES's work agenda consists of three major areas: events, projects, and publications. The events calendar consists of conferences sponsored by the organization itself, the most important of which are its biennial Solar World Congress and a number of regional conferences; conferences in which it is a cosponsor or supporting group, such as the IEA Solar Heating and Cooling Conference and the Solar Energy Technology in Development Cooperation meeting; and webinars on topics such as Solar Food Processing for Income Generation and Solar Driven Thermochemical Production of Sustainable Fuels. ISES projects are ongoing activities that the organization describes as the "life blood" of its existence. Among recent project topics have been research on renewable energy deployment in Brazil, China, and South Africa; development of an online guide for users of solar datasets; and creation of the ISES Solar Education Exchange, a mechanism by which students around the world are able to interact with each other to advance the goal of moving toward a solar energy future.

Among the organization's major publications are its official monthly journal, *Solar Energy*, which contains scholarly papers on a variety of solar topics, such as solar energy measurement, research, development, application, and policy. *Renewable Energy Focus* is a publication of more general interest for the general reader, with articles on all aspects of renewable energy. The *ISES Membership Newsletter* deals with topics of more specific interest to members of the organization, such as news of the organization as a whole, section news, partner news, member news, and corporate member news. ISES president Dr. David Renné also maintains his own SunBurst blog, with news about the organization's activities and his own personal experiences and opinions. The association also publishes occasional white papers and reports on a variety of topics related to solar energy research and development. In addition to publications from the international organization itself, some sections also produce their own publications, such as *Solar Today*, produced by the American Solar Energy Society; *Sonnenenergie* (*Solar Energy*), the official magazine of the Deutsche Gesellschaft für Sonnenenergie e. V.; and *Solar Progress*, the official journal of the Australian Solar Council.

Augustin Mouchot (1825–1912)

Mouchot was one of the first individuals of the modern era to promote the notion that fossil fuels, so important a part of the Industrial Revolution, could inevitably not last forever and that inventors and industrialists would have to think about possible alternative sources of energy. He was especially interested in the possibility of using solar energy as a substitute for fossil fuels when the time came that coal, oil, and natural gas were no longer available to operate machines and power most forms of transportation. Mouchot spent a significant part of his life developing devices that could operate on solar power and for a time was internationally recognized for his contribution to the field. The last years of his life were marked, however, by personal disappointment in the world's decision to gamble its energy future on the fossil fuel and by financial ruin and social rejection.

Augustin Mouchot was born in the small town of Semur-en-Auxois in the department of Côte-d'Or in eastern France on April 7, 1825. For the first half of his life, he seemed destined to lead a somewhat undistinguished life as a teacher of mathematics at the primary and secondary level. His first job was as a teacher at the primary school in the Morvan district in central Burgundy (from 1845 to 1849) before taking a similar position in the city of Dijon, capital of the Côte-d'Or. By 1852 Mouchot had earned a degree in mathematics and, a year later, a second degree in physical science, qualifying him to teach at the secondary level. In 1853, then, he accepted a position at the secondary school at Alençon, in Normandy, where he remained for almost a decade, before moving on to similar positions in Rennes, and finally at the Lycée de Tours (1864–1871).

As early as 1860, Mouchot also began to spend his spare time on his true field of interest, solar energy. He began to conduct experiments to find ways of capturing and concentrating solar radiation as a means for producing useful energy. His work was a continuation of earlier studies by inventors such as of Horace-Bénédict de Saussure and Claude Pouillet, who had designed "solar ovens" that could be used to boil water using the Sun's rays. The basic principle behind Mouchot's devices was a jar filled with water with a glass cover. The container was lined with a black material that absorbed solar radiation to a point at which the heat within the container was sufficient to boil the water. The steam produced by this device was sufficient to operate a small steam engine. The greater part of Mouchot's research was devoted to finding the most efficient way of capturing and focusing solar radiation on the water-filled container so as to produce the maximum amount of steam from the system.

Mouchot's work achieved its acme in 1869 when he published the book for which he is best known, *La Chaleur solaire et ses Applications industrielles* (*Solar Heat and Its Industrial Applications*), which summarized the results of his extensive series of experiments on solar energy. In the same year, Mouchot unveiled his penultimate solar machine in Paris, the largest such device to have been built to that date. Unfortunately, the machine disappeared during the Franco-Prussian War in 1871, although a diagram of Mouchot's invention still exists (see http://landartgenerator.org/blagi/archives/2004). The 1869 device was sufficiently impressive, however, that the French government agreed to provide some financial support for Mouchot's continued experimentation on solar energy. One such grant allowed Mouchot to continue his research on solar devices in Algeria, where the amount of sunlight available for solar engines was much greater than it was in France itself.

The largest and most impressive of Mouchot's many inventions was a solar engine built for the Universal Exhibition held in Paris in 1878. The device had a reflecting mirror about 4 meters (13 feet) in diameter and a boiler with a capacity of 80 liters (21 gallons). It could be used both for boiling water and, strangely enough perhaps, for making ice. Mouchot was awarded a gold medal for outstanding achievement by the exhibition.

Mouchot's success in designing ever more impressive solar engines might have seemed an encouraging direction for his future career. The problem was, however, that fossil fuels continued to be available in ever-increasing amounts. The notion that the world would someday run out of coal, oil, and natural gas might have made sense from a theoretical standpoint, but there appeared to be no imminent threat of that kind at the end of the nineteenth century. Indeed, the fuels needed to operate the world's machinery were still available at only a few dollars per Btu (British thermal unit), well within reach of almost any industry. By contrast, solar energy, while a fascinating alternative that captured the imagination of exhibition-goers, was still many times more expensive to produce and use than were the fossil fuels.

By 1879, Mouchot could see the handwriting on the wall for solar energy. He decided to give up his research on solar engines and to return to teaching, a plan that he was unable to follow, however, because of a bacterial infection that he developed and that prevented him from returning to work. He was able to receive a modest disability pension from the French government, but it was inadequate to meet his personal needs and he slowly fell into poverty. His life was made even more difficult by the loss of his wife, who was moved to an asylum because of mental disorders. By the time he died in Paris on October 4, 1912, he was so ill that he was unable even to go to the post office to pick up the modest pension on which he was surviving.

Elon Musk (1971–)

Musk is a billionaire engineer, inventor, and entrepreneur who may be best known as the product architect for the Tesla motor company and as chief executive officer and chief technical officer of the space program called SpaceX. He is also responsible for developing the plans for a new solar energy company called SolarCity, which, in 2015, described itself as "America's #1 full-service energy provider." SolarCity began operations in 2006 under the leadership of Musk's cousins, Lyndon Rive and Peter Rive, As of 2015, SolarCity was operating in 18 states (Arizona, California, Colorado, Connecticut, Delaware, Hawaii, Maryland, Massachusetts, Nevada, New Hampshire, New Jersey, New Mexico, New York, Oregon, Pennsylvania, Rhode Island, Texas, Washington) and the District of Columbia. The system by which SolarCity provides services to individual customers consists of three simple steps, beginning with a personal consultation about the electrical needs of a particular building. The consultation is followed up by design of a solar system intended for that specific building. Finally, SolarCity workers install the system on the building and the new system is ready to operate. SolarCity signed its 100,000th customer in the first quarter of 2014, and CEO Lyndon Rive announced at that point that he was aiming to have one million customers by the end of 2018.

In 2012, Musk announced that SolarCity would partner with Tesla Motors in a program that would use Tesla car batteries as storage systems for SolarCity roof panels. In 2015, the SolarCity-Tesla partnership went one step farther with the announcement of a new type of grid, called a microgrid, designed for smaller geographical region that traditional electrical grids. Called GridLogic, the system was being targeted for college campuses, military bases, remote communities, and small municipalities.

Elon Reeve Musk was born in Pretoria, South Africa, on June 28, 1971. His father is Errol Musk, a South African–born British electrical and mechanical engineer and part owner of an emerald mine. His mother is Maye Musk, a registered dietician, popular spokesperson for wellness programs, and long-time model. She is said to be still in demand for modeling work at the age of 60+. Musk's parents divorced in 1980, and he chose to live most of his time with his father, even though, he once told an interviewer from *The New Yorker*, he is "not a fun guy to be around."

Musk had an active curiosity about both science and economics at an early age. When he was only 12, for example, he invented a video game called Blast Star (or Blastar) that combined the elements of two other popular video games, Asteroids and Space Invaders. He sold his invention for $500.

Musk attended school at the Waterkloof House Preparatory School and the Pretoria Boys High School, from which he graduated in 1988. He then decided to move to Canada, where he thought opportunities for advancement would be greater than they were in South Africa.

He was granted Canadian citizenship because of his mother's own status as a Canadian citizen.

In 1990, Musk matriculated at Queen's University in Kingston, Ontario. After two years, he decided to transfer to the University of Pennsylvania, from which he received his BS degree in physics and, concurrently, his BA in economics from the Wharton School of Business in 1994. He then moved to California with plans to pursue PhD in applied physics at Stanford University. He stayed in the program for only two days, however, before deciding that he was really more interested in getting on with his own inventions in the field of electronics and his own entrepreneurial plans for creating new businesses.

Musk's first endeavor along those lines with the creation of a web software company called Zip2 with his brother Kimball. The company developed city guides for newspapers such as *The New York Times* and the *Chicago Tribune*. Four years later, the company was sold to Compaq for $307 million in cash and $34 million in stock options, earning Musk a profit of $22 million.

Musk then immediately turned his attention to another venture, an email payment company he called X.com. Two years later, in 2001, X.com merged with a similar company,

Confinity, which included an online funds transfer system called PayPal. Only a year later, the combined operation was sold to eBay for $1.5 billion in stock, earning Musk a profit of $165 million.

Again, Musk moved on to another business venture as soon as the eBay deal was closed, a company focused on the design and manufacture of space vehicles called Space Exploration Technologies, or SpaceX. Musk continues to serve as the chief executive officer and chief technology officer of the corporation. In May 2012, the company's SpaceX Dragon vehicle became the first commercial rocket to fly to and dock with the International Space Station. Musk and a number of other experts in the field claim that this accomplishment marks the beginning of transitioning space flight programs for government to private operations.

In 2010, Musk was one of a group of five investors who formed Tesla Motors, a company created to design and build electric cars. The company had been created in 2003 but had not achieved the financial stability to actually produce cars until almost a decade later. Then, in 2008, the company began general production of its first commercial product, a fully electric sports car called the Tesla Roadster. In 2012, the company released its second vehicle, a sedan called Model S. The Model S received some measure of fame in 2013 when it became the best-selling car in Norway, the first time in history that an electric car had achieved that rank in any country in the world.

Although still a relatively young man, Musk has already received a number of awards and honors, including the Gold Space Medal of the Fédération Aéronautique Internationale; Heinlein Prize for Advances in Space Commercialization; George Low Prize of the American Institute of Aeronautics and Astronautics; Von Braun Trophy of the National Space Society; National Conservation Achievement Award of the National Wildlife Federation (for his work with Tesla and SolarCity); and Entrepreneur of the Year Award from Inc. magazine.

National Renewable Energy Laboratory

The National Renewable Energy Laboratory (NREL) was established by the Solar Energy Research Development and Demonstration Act of 1974 as the Solar Energy Research Institute (SERI). The NREL began operations in July 1977 and was designated a national laboratory of the U.S. Department of Energy (DOE) in September 1991. The NREL's main campus is located in Golden, Colorado. The agency also has administrative offices in Washington, D.C.

NREL conducts research on all forms of renewable energy, including hydrogen fuel cell and related technologies; bioenergy; and wind, solar, and geothermal technologies. Some of the research units at which this research occurs are the National Center for Photovoltaics, the National Bioenergy Center, the Thermochemical Users Facility, the Battery Test Facility, the Thermal Conversion Facility, the Ethanol Process Development Unit, the Solar Radiation Research Laboratory, the Photovoltaics Outdoor Test Facility, the Solar Furnace, and the Renewable Fuels and Lubricants (ReFUEL) Research Laboratory, all located at the Golden campus. The agency's overall objective is to develop cleaner, more reliable, and more affordable energy options that will reduce air pollution and greenhouse gas emissions, strengthen the nation's energy security, improve electric grid operations, boost local economic development, and increase energy and economic efficiency. Although the agency's primary emphasis is on America's energy equation, it also has a commitment to improving energy options throughout the world and has contributed to renewable energy projects in rural communities in Africa, Asia, and South America.

The work of the NREL falls into four general categories: energy analysis, science and technology, technology transfer, and applying technologies. The agency's energy analysis components involves the analysis of technical systems, markets, policies, and levels of sustainability as well as the development of models and tools for the study of renewable resources. These studies have produced a large number of publications, perhaps most useful of which for the general public are the annual Renewable Energy Data Book and Renewable Energy Market Data, both of which are available for download on the agency's website (http://www.nrel.gov/analysis/).

The NREL science and technology component includes research studies undertaken by the agency on the major forms of renewable energy and has resulted in reports on topics such as hydrogen gas turbine development, advanced vehicle drive systems, cellulose digestion in the production of bioenergy, data availability for home energy retrofits, bio-hybrid fuel cell development, material development for improved solar cells, and analysis of current offshore wind energy developments. Examples of some of the specific research areas undertaken by the NREL in just one area, advanced vehicles and fuels research, are electrical vehicle grid integration, fleet test and evaluation, fuel technology impacts, petroleum-based fuels, power electronics, nonpetroleum based fuels, regulatory support, and vehicle systems analysis.

The technology transfer component of NREL's work involves cooperative efforts between the agency and private corporations for the implementation of research conducted at the NREL. This cooperation involves work conducted jointly between the agency and private corporations, the licensing to corporations of discoveries and advances made at the NREL, and work carried out by the NREL for energy corporations. One of the "success stories" touted by the NREL in its technology transfer program has been the development of a new type of material for use in the construction of solar cells. The new material makes use of a new kind of metal foil called rolling assisted biaxially textured substrates (RABiTS), originally developed at the NREL on which silicon crystals are deposited out of the gas phase in a process developed by the Ampulse Corporation. NREL and Ampulse researchers hope that their new material will provide a breakthrough in the process by which solar cells can be produced with a significant reduction in the cost of production of cells.

The applying technology component of NREL's mission involves the agency's assistance to businesses and institutions, federal government agencies and facilities, state and local governmental agencies, tribal communities, and international and regional governments and institutions in implementing and adapting NREL discoveries and advances to each specific entity's special needs and conditions. As an example, the NREL assumed a major role in the rebuilding of the town of Greensburg, Kansas, after the community had been essentially destroyed by a tornado that hit the area in May 2007. The NREL was able to make suggestions for reconstruction that eventually reduced the community's carbon dioxide emissions by 36 percent over pre-tornado levels.

The NREL has an extensive collection of publications, many of which are available online. They include popular topics such as *Consumer's Guide: Get Your Power from the Sun* (brochure); *Elements of an Energy-Efficient House* (fact sheet); *Borrower's Guide to Financing Solar Energy Systems: A Federal Overview* (booklet); *Junior Solar Sprint: An Introduction to Building a Model Solar Car* (instructional booklet); *Wind Energy Benefits* (fact sheet); *Look Back at the U.S. Department of Energy's Aquatic Species Program: Biodiesel from Algae; Close-Out Report* (report); *Lessons Learned from Existing Biomass Power Plants* (booklet); *Axial Flux, Modular, Permanent-Magnet Generator with a Toroidal Winding for Wind Turbine Applications* (technical report); *International Performance Measurement and Verification Protocol: Concepts and Options for Determining Energy and Water Savings, Volume I* (protocol); and *Concentrating Solar Power: Energy from Mirrors* (PDF file).

The NREL also conducts an energy leadership program called the Executive Energy Leadership Academy. The program is focused on providing individuals with a nontechnical explanation of the opportunities available for the use of renewable energy options in a variety of government and business settings. Two program options are available, one consisting of five two-day sessions completed over a period of five months, and the other consisting of one three-day session completed within a one-month period.

Russell S. Ohl (1898–1987)

Ohl was an American electrical engineer who is generally credited as having invented the earliest version of the modern solar cell. In 1937, he discovered that the presence of impurities in an otherwise very pure silicon crystal was responsible for the flow of electrons from one portion of the crystal to a different part of the crystal when that crystal was exposed to light. In 1946 he was granted a patent (US Patent 2402662) for the invention of a "light-sensitive electronic device" that operated on the principle he discovered nearly a decade earlier.

Russell Shoemaker Ohl was born in Macungie, near Allentown, Pennsylvania, on January 30, 1898. He was apparently a very bright child who, as he told one interviewer for the Center for the History of Electrical Engineering (CHEE), started school at the age of five and then began "skipping a couple of grades here and there." He graduated from high school at the age of 16, after completing 11th grade, before matriculating at the Keystone State Normal School, which is now part of the University of Pennsylvania system. At Keystone he enrolled in the courses that most appealed to him, in the fields of chemistry and engineering. In one of those courses he saw a wireless radio receiver, the first such device he had ever seen. He was able to pick up radio communications from a ship in the Atlantic Ocean that was under attack by a German submarine and was fascinated at the radio's ability to carry messages over such long distances. He explained to the CHEE interviewer that "that was when the radio bug bit me. I never got over it."

After completing his courses in chemistry and engineering at Keystone, Ohl enrolled at Pennsylvania State College, now Pennsylvania State University. His time there was intensive, with no breaks for vacations at any time of the year, primarily because of the pressures exerted by World War I. After receiving his degree in electrical engineering from Penn State in 1919, he enlisted in the U.S. Army, where he was assigned to the Signal Corps. He continued his training for the corps back at Penn State, from which he had just graduated.

At the war's completion, Ohl found himself somewhat at loose ends, not knowing exactly what to do next. Eventually he took a research job with the Electric Storage Battery Company, where he was assigned the task of designing a battery for airplane telephone transmitters. He did not find the work very interesting and so accepted an offer to move to the Westinghouse Lamp Company, where he was also bored with his assignment to work on lamp development. He said that he became more interested in working with vacuum tubes, a project on which he spent all of his free time.

After a short time with Westinghouse, Ohl decided on another move, this time to the University of Colorado, where he took a job as an instructor. Again, the job was not exactly what he was looking for, and he remained at Colorado for only a year. Although he was offered a contract to continue working at Colorado, Ohl instead accepted a chance to join the research staff at the American Telephone & Telegraph Company (AT&T) in New York City. During his five years at AT&T, Ohl finally had an opportunity to spend his time working on the subject in which he was most interested: radio technology. He told his CHEE interviewer that, while he was at the company, he did "almost everything there was to do regarding radio," including the development of quartz crystals for rectification in radio systems, the line of work that was eventually to lead to the discovery for which he is best known today.

In 1927, Ohl moved once again, this time to AT&T's primary research arm, Bell Laboratories (better known simply as Bell Labs) in Murray Hill, New Jersey, where he remained for the rest of his professional career. It was at Bell Labs that Ohl designed what is now generally regarded as the first solar cell. The patent he received for that invention was one of 82 U.S. patents and 40 international patents for a variety of radio-related technical innovations. Ohl once described his years at Bell as being satisfying in some ways, because of the many fields of research in which he was able to work, but also a bit frustrating in that his superiors were constantly asking him to leave that work and focus on more immediate practical problems of radio design and operation.

Ohl retired from Bell in 1958 and moved to California, where he spent the rest of his life. He died in Vista, California, on March 20, 1987.

Stanford R. Ovshinsky (1922–2012)

The two terms most commonly associated with Ovshinsky have been "prolific" and "self-taught." Without the benefit of a college education, Ovshinsky invented a number of important electronic devices for which he was awarded more than 400 patents over a 50-year period of work. Many of his inventions were in the field of solar energy, the most important of which was the continuous web multi-junction flexible thin-film photovoltaic panel, usually referred to simply as the amorphous thin-film solar panel. This device can be produced, as the name suggests, in the form of a lightweight, flexible sheet of solar cells that is now the most widely used material for solar roof arrays.

Ovshinsky's invention of thin-film cells was based on an earlier hunch about the type of silicon that would work best in solar cells. For many years in the early history of solar technology, most people believed that only ultrapure crystals of silicon would function properly in a solar cell. In the 1950s, Ovshinsky suggested that such might not be the case and that, instead, amorphous (non-crystalline) silicon might work as well in solar cells and would have the added advantage of being much less expensive to make. Most experts rejected Ovshinsky's suggestion, especially as it came from someone without a college education in the field of semiconductors or solar technology. As it developed, Ovshinsky was correct, and many types of solar cells today are made from amorphous, rather than crystalline, silicon. Because of his many contributions to the field, Ovshinsky has sometimes been called the father of modern solar energy (or technology).

Stanford Robert Ovshinsky was born on November 24, 1922, in Akron, Ohio. His father was Benjamin Ovshinsky, an immigrant from Lithuania, and his mother was Bertha Munitz Ovshinsky, an immigrant from Belarus. Benjamin Ovshinsky was a scrap metal dealer in Akron, an occupation that made it possible for him to get his son a job as a lathe operator while Stanford was still in high school. During World War II, Ovshinsky was exempted from military service because of his asthma, so he moved instead to Phoenix, Arizona, in order to work in a Goodyear aircraft factory. At the war's end, he moved back to Akron and opened his own machine shop, where he patented his first invention, a new type of lathe. He was so successful in his Akron business that he was able to sell his shop to the New Britain (Connecticut) Machine Company in 1950. After operating the shop for New Britain for two years, he moved to Detroit to take a job with the Hupp Motor Company, then producer of the popular Hupmobile car.

While working in Detroit, Ovshinsky became especially interested in two areas that were eventually to inform much of his research in later years: intelligent machines and neurophysiology. He recognized early on that machines need not necessarily be the dumb, plodding devices capable simply of repeating the same process over and over again without changing their own basic "knowledge" of a procedure or altering their own behavior. Instead, he suggested that machines could be made capable of learning about a procedure as they carried it out so that it could perform that function at later times more effectively and more efficiently. At one point, he opened a shop called General Automation in Detroit with his younger brother Herbert to carry out research on this idea. At the shop, the two Ovshinsky brothers developed a semiconductor switch that operated on the same principle as a neuron, a device they called the Ovitron. The Ovitron was capable of converting a small change in voltage to a large variation in current and was tested by the U.S. Air Force for possible use. It was rejected not because it didn't work properly, but because it was the subject of possible catastrophic accidents. (It contained concentrated sulfuric acid, which could spill during an accident.)

In 1960, Ovshinsky and his second wife, the former Iris L. Miroy, founded a company they called Energy Conversion Laboratories (later called Energy Conversion Devices) in Rochester, Michigan. The purpose of the company was to commercialize some of Ovshinsky's inventions, although it proved to be singularly unsuccessful in achieving that objective. The company eventually reached its 35th year of operation without ever making a profit. The Ovshinskys were able to survive largely because of their ability to convince potential investors and partners over and over again of the ultimate value of their inventions, and the bonanzas that would result when commercialization actually began to occur (which it never did). While fending off bankruptcy with one hand, however, Ovshinsky remained uniquely skilled at turning out new inventions one after the other, until the final years of his life.

Iris Ovshinsky died in 2006, and a year later Ovshinsky remarried and formed another new company, Ovshinsky Innovation LLC, focused more on basic research and the development of yet new innovations in the field of energy. Ovshinsky died at his home in Bloomfield Hills, Michigan, on October 17, 2012, at the age of 89. Among the many honors and awards he received during his lifetime were the 2005 Innovation Award for Energy and the Environment from The Economist magazine; the Hoyt Clarke Hottel Award of the American Solar Energy Society; Karl W. Böer Solar Energy Medal of Merit of the University of Delaware Institute of Energy Conversion; Sir William Grove Award of the International Association for Hydrogen Energy; Walston Chubb Award for Innovation given by

Sigma Xi research society; Lifetime Achievement Award of the Engineering Society of Detroit; Environmental Hall of Fame 2008 Award, Solar Thin Film Category; and Thomas Midgley Award of the Detroit Section of the American Chemical Society. Ovshinsky was inducted into the 2005 Solar Hall of Fame and was awarded honorary doctorates from Kettering University, Flint, Michigan; University of Michigan; Wayne State University; Illinois Institute of Technology; Ovidius University, Constanta, Romania; New York Institute of Technology; and Kean University, Union, New Jersey.

Frank Shuman (1862–1918)

Shuman was a visionary inventor who saw the potential of solar energy as a source of power for human societies nearly a century before that concept began to develop in the modern world. He was granted more than 60 patents between 1892 and 1922 for methods for making a variety of types of sheet glass, producing wire-embedded glass products, producing various tools used in the manufacture of glass, forming concrete piles and casings, extracting grease and potash from wool, making a new kind of danger signal, design of a submarine along with its operation, and the manufacture of steam engines that operate on solar power.

Shuman is probably best known for the design of the world's first solar thermal power station, built in Maadi, Egypt, between 1912 and 1913. The plant used 60-meter-long parabolic troughs to capture and concentrate solar energy a half century before that technology became widely used in the United States and other parts of the developed world. The plant produced about 60 horsepower of power and was used to pump water from the Nile River to cotton fields in the region. Shuman has been quoted as saying that the plant "proved the commercial profit of sun power in the tropics and has more particularly proved that after our stores of oil and coal are exhausted the human race can receive unlimited power from the rays of the sun." (Nick Blair. "Frank Shuman—The Father of Solar Power." Solar.Pub. http://solar.pub/frank-shuman-father-of-solar-power/. Accessed on July 22, 2015.) The plan remained in operation for only about two years but was then torn down to provide metal parts for the British munitions industry ramping up for World War I.

Frank Shuman was born in Brooklyn, New York, on January 23, 1862. He had little formal education but possessed an avid interest in science and taught himself the rudiments needed to pursue his interest in inventing. At the age of 18, he began his first paid job with the aniline dye company of Victor G. Bloede in West Virginia. In his spare time, he began working on a number of inventions, the first of which he patented was a form of plate glass embedded with wire to improve the strength and safety of the material then widely used for skylights. The success of that invention led Shuman's uncle, Frank Schumann, to convince Shuman to move to Philadelphia, where he (Schumann) was president of the Tacony Iron & Metal Works. Shuman's first assignment at his uncle's plant was to design an electroplating process that could be used on the statue of William Penn that was being made for the top of Philadelphia's city hall. His solution to this problem proved to be so successful that it was described in article about the process in the prestigious journal *Scientific American* within weeks of the statue's installation. It also brought Shuman sufficient financial success to allow him to open his own business in the Tacony section of Philadelphia, Shuman's American Wire Glass Manufacturing Company, in 1892.

Although most of his patents dealt with innovations in the production of glass and concrete structures, Shuman long had a passion for the importance of solar energy as a way of meeting human energy needs. In a 1914 article in the *Scientific American*, for example, he wrote that "[o]ne thing I feel sure of is that the human race must finally utilize direct sun power or revert to barbarism." That expression of hope and faith came fast upon his success at the Maadi solar facility. But it reflected an interest in solar energy that went back almost two decades. As early as 1897, for example, Shuman was experimenting with the design of simple solar engines made of glass-topped black boxes filled with ether. Shuman chose ether for his working fluid because it boils at a much lower temperature than does water. One of his early models operated successfully for low-power purposes at a pond beside his home in Tacony.

By 1908, Shuman was ready to commit to a larger investment in solar power. In that year he founded the Sun Power Company with plans for building larger and more powerful solar engines that would be capable of competing with fossil-fueled power stations. His design for these engines involved larger boxes with larger glass tops with water rather than ether as the working fluid. He also designed a new type of low-pressure steam turbine since even the best of his solar engines did not produce enough power to operate a traditional steam engine powered by coal. He obtained a patent for this new and more powerful engine in 1912.

Shuman was unsuccessful, however, in finding investors in the United States to support his new enterprise. So he turned to Great Britain, a nation with extensive business holdings in its colonies around the world, many in sun-soaked regions such as Egypt and India. The Maadi plant was made possible by the financial support of a group of British investors who saw Shuman's idea as a pregnant approach to producing energy in such regions. Additional plans for building other solar power plants in areas such as the Sahara Desert had to be put on hold, however, because of the advent of World War I. By the time that war was over, Shuman had died of a heart attack seven months earlier, on April 28, 1918, at his home in Tacony, the world remained committed to fossil fuel–produced energy, and the promise of solar power was delayed by another half century.

Willoughby Smith (1828–1891)

Smith is important in the history of solar energy because of his accidental discovery of the photoconductive properties of the element selenium. That discovery made possible the invention of some of the earliest types of photovoltaic solar cells, devices in which light is converted to electricity by means of an intermediary material, selenium being one of the most efficient and most widely used.

Willoughby Smith was born at Great Yarmouth, England, on April 16, 1828. At the age of 20, he entered the employment of the Gutta Percha Company of London. The company manufactured a rubber-like material called gutta percha that was used as a covering for electrical cables, such as those used in telegraphic systems. Smith's original assignment with the company was to study the ways in which gutta percha could be used for such purposes. In 1849, he became involved in the laying of a telegraphic line under the English Channel between Dover and Calais, consisting of a copper wire encased in a gutta percha wrapping. He made his professional name on the project when he developed a way of joining adjacent segments of the line in such a way as to prevent water from seeping into the connection and interrupting the transmitted signal.

Smith's experience on the Channel line set the stage for the type of work he was to do for the rest of his life: designing, constructing, laying, and testing underground and underwater cables. He was involved in such major projects as the laying of the first telegraphic cable in the Mediterranean Sea between La Spezia and Corsica, Corsica and Sardinia, and Sardinia and Bona. In 1858, he became involved in the most ambitious cable-laying project of all, a transatlantic connection between Ireland and Newfoundland. One of the challenges with which Smith had to deal on this project was the testing of the cable as it was being laid, to make sure that no breaks had occurred in the process. The test he employed required the use of a substance that conducts electricity, but not as well as a good conductor such as copper or aluminum. Smith decided to use the element selenium for this process, an element now known to be a semiconductor, a substance that does conduct electricity, but only very poorly. To Smith's considerable surprise, his selenium test bars behaved in a very erratic fashion. At night or in otherwise darkened conditions, selenium behaved as expected, conducting a very weak electrical signal. In daylight or other bright light, however, selenium suddenly became a very good conductor, similar to the behavior of many metallic elements.

When the opportunity presented itself, Smith designed an experiment to learn more about the effect he had observed in his cable work. On February 20, 1873, he submitted a letter to the journal *Nature* in which he summarized the results of his studies. He explained that he placed a bar of selenium in a box with a sliding lid and then connected both ends of the bar to an electric meter. He found that "[w]hen the bars were fixed in a box with a sliding cover, so as to exclude all light, their resistance was at its highest, and remained very constant, fulfilling ail the conditions necessary to my requirements *[for the cable project]*; but immediately the cover of the box was removed the conductivity increased from 15 to 100 per cent, according to the intensity of the light falling on the bar. Merely intercepting the light by passing the hand before an ordinary gas-burner, placed several feet from the bar," he continued, "increased the resistance from 15 to 20 per cent."

Smith spent the rest of his working life with the Gutta Percha Company and its successor, the Telegraph Construction and Maintenance Company, where he was eventually appointed chief electrician. He continued to be involved in the laying and testing of underwater cables in the Atlantic Ocean until the strain became too great, and he retired from active participation in projects in order to recover his health. Smith died at his home in Eastbourne, England, on July 17, 1891.

Solar Electric Power Association

The Solar Electric Power Association (SEPA) was founded in 1992 as the Utility PhotoVoltaic Group (UPVG). The organization was originally created for the purpose of promoting investment in large-scale photovoltaic facilities with the goal of reducing the cost of solar PV electricity through economies of scale. Founders of the original organization were the Edison Electric Institute, American Public Power Association, National Rural Electric Cooperative Association, and Electric Power Research Institute. The original membership list originally consisted of 13 utilities, although that list grew rapidly and reached 76 companies only four years after the founding date. As a result of the changing structure of the solar energy market at the end of the twentieth century, UPVG changed its name in October 2000 to its present title of Solar Electric Power Association. Today, SEPA has more than 500 members distributed among three major categories: utilities, corporate, and government, non-profit, and education. The utilities groups consist of investor-owned utilities, public power and municipal utilities, rural electric cooperatives, generation and transmission cooperatives, irrigation and water districts, and the Federal Power Marketing Administration. Corporate members include business and professional services; project developers, installers, and distributors; manufacturers; independent power producers; power marketers; and retail electricity providers. The final category includes governmental agencies at all levels, federal, state, and local; colleges and universities; and nonprofit organizations.

One of the major focuses of SEPA's work is to conduct research on electricity derived from solar power so as to educate professionals in the field and the general public so as to create a greater interest in and support of solar electric power. The organization provides an extensive variety of tools on its websites that are available only to members to assist in their planning and activities. Some examples are the Utility Solar Database, which provides an extensive array of information about nearly every aspect of a company's structure and activities, such as financing and customer rates (available only to members). Another popular publication is the association's annual "SEPA Utility Solar Rankings and Top 10," a listing of the nation's most successful solar electricity companies, along with up-to-date information on current trends in the development and utilization of solar energy. Other online tools can be used by members to find research on solar electricity technology and funding, business models, and solar projects. Personalized research is also available on specific questions of interest to members through SEPA's staff researchers.

Another large field of SEPA activities involves the facilitation of information dissemination and networking among member organizations. One device developed for this purpose is called SEPA Connect, an online forum through which members are able to connect and communicate with each other. Members are also invited to participate in virtual online working groups on issues of particular interest to those involved in the solar electricity industry. SEPA also sponsors and cosponsors a variety of annual and onetime conferences, webinars, and other meetings through which members can establish contact with each other and receive updates on the status of solar electricity power generation in the United States.

The three main topics around which SEPA organizes its programs are utility solar business models, policy, and technology. The first topic is designed to help members understand and make use of new ways of doing business that reflect changes in governmental policy, technological developments, and other factors affecting the solar electricity industry. The area of policy involves an exploration and understanding of all phases of policy making that have an effect on the field of solar electricity generation, such as actions taken by the U.S. Congress, new rules and regulations created by federal agencies, new programs for incentives and financing in solar power, planning for and adoption of interconnection standards, net energy metering policies and practices, issues surrounding permitting and siting, policy innovations, rate design, Renewable Portfolio Standards (RPS), and state actions. Technology is a fundamental feature of SEPA action, of course, since it lies at the base as to what companies can and cannot do in the field of solar electricity generation.

Three of SEPA's major events of the year are its annual Utility Solar Conference, Solar Power International (cosponsored with the Solar Energy Industries Association), and its fact-finding missions. The last of these consists of a group visit to some facility or location to gain information about developments in the solar electricity industry. As an example, a September 2014 fact-finding mission to Germany was designed to help participants gain a better understanding of the transition being made from conventional fossil fuels to solar electricity in that nation.

A feature of special interest among SEPA's programs is its internship program, in which college juniors or seniors or graduate students have an opportunity to work in one of three fields within the solar electricity industry: research, member relations, and marketing and communications. Internship programs run for about three months and operate in summer, fall, and spring sessions. For more information about these programs, refer to SEPA's home page at https://www.solarelectricpower.org/about-sepa/internship-information.aspx.

Solar Energy Industries Association

The Solar Energy Industries Association (SEIA) was founded on January 24, 1974, by representatives of five solar energy companies—Jim Edison, Peter Glazer, James Ince, Yahya Safdari, and Sam Taylor—met (according to association records) "in the noisy basement of the Washington Hilton" to discuss the formation of a professional association of organizations interested in solar energy. The original mission of the founders was to create "a broad-based trade association supporting prompt, orderly, widespread and open growth of solar energy resources now."

The need for such an organization could not have come at a more obvious time. The early 1970s were a period in which interest in solar energy was at its highest point in modern history. Legislation passed in the period included the Solar Heating and Cooling Demonstration Act of 1974, the Energy Reorganization Act of 1974, and the Solar Energy Research Development, and Demonstration Act of 1974, all of which were designed to expand and increase the nation's commitment to and investment in the development of solar energy. The opportunities for solar industries to take advantage of the federal government's new largesse in the investment in solar energy had probably never been at such a peak.

The irony facing the new association, however, was that solar industries in the United States were still in their very earliest stage of development. The question raised by some of the earliest members of SEIA was whether creation of the new organization might be premature. As Sheldon Butt, the first president of SEIA, asked at the time, "Shouldn't we have an industry before we can have an industry association?" Members of the new association were obviously not too troubled by such niceties, however, as they charged ahead with the formation and functioning of the new organization.

SEIA currently consists of about 650 corporate members in essentially every field of solar energy, including contracting and installation (about 200 members); manufacture and supply (128); project development (67); consulting (40); finances (39); distribution (37); service providers (35); engineering, procurement, and construction (24); legal services (16), not–for–profit services (14); communication and marketing (11); education (5); architectural and engineering (4); research (4); and other categories (about 20). The association offers five levels of membership, each with its own set of benefits: basic, kilowatt, megawatt, gigawatt, and terawatt. Benefits associated with membership include access to online information and services, discounts on various association publications, participation in committees at each level of membership, free access to SEIA webinars, free registration to SEIA conferences and other meetings, and ability to run for certain SEIA offices.

SEIA also maintains forma working relationships with 15 state and regional chapters, such as AriSEIA (Arizona Solar Energy Industries Association); COSEIA (Colorado Solar Energy Industries Association); OSEIA (Oregon Solar Energy Industries Association); and GSREIA (Gulf States Renewable Energy Industries Association), representing industries in Alabama, Louisiana, and Mississippi. This relationship developed as a result of SEIA's 2012 merger with the Solar Alliance, an advocacy group working to develop solar policies at the state level. Links to individual state chapters are available on the SEIA website at http://www.seia.org/about/seia/official-state-chapters.

Much of the association's work is carried out by membership-level-related committees, such as (for basic membership level): corporate social responsibility, codes and standards, installer safety, and wholesale distribution generation; (for kilowatt level): finance, public relations, tax and accounting, and trade and competitiveness; (for megawatt level): federal affairs, state and regional committees, and Solar Beltway Task Force; (for gigawatt and megawatt levels): federal policy, post-2016 task force, net metering task force, and governance task force.

SEIA claims to represent every aspect of the solar industry, "from the small-business owners to the multi-national companies, from the installers on the roof to the engineers in the lab," and that claim is validated by the wide range of issues and policies it had chosen to work on. Those issues and policies fall into eight general categories: distributed solar, power plant development, international trade, environment, finance and tax, renewable energy deployment, health and safety, and solar technology. Each of these categories is subdivided, in turn, into more manageable topics such as, in the case of distributed solar, topics such as shared renewables, interconnection standards, local permitting, net metering, property-assessed clean energy, rebates and incentives, solar access rights, and utility rate schedule. At the simplest level, SEIA provides background information, additional resources, useful fact sheet, and other materials related to each of these topics.

An important feature of SEIA's work is its program of research on important issues and policies related to solar energy. The association publishes research reports, fact sheets, white papers, and other materials that summarize the results of its research on such topics. Examples of some recent publications are the organization's 2014 "U.S. Solar Market Insight;" a report on "Cutting Carbon Emissions under EPA Section 111(d);"

the "National Solar Database," which aggregates and reports on detailed information about solar energy companies operating in the United States; a 2015 fact sheet on "Major Solar Projects in the United States," and the "National Solar Survey" that reports on voter attitudes about solar energy in the United States.

In working to achieve its goals, SEIA sponsors and takes part in a number of national, regional, and state conferences; webinars; and other types of meetings and educational events. Examples of recent events are an association finance and tax seminar held twice a year; PV America, a regionally focused tradeshow of solar energy–related technology and materials; Solar Power International, claimed by SEIA to be "the most powerful solar event in the U.S.," along with regional conferences of a similar nature; and webinars on a variety of solar energy–related topics.

Robert Stirling (1790–1878)

Stirling is best remembered today as the inventor of a heat engine that carries his name, the Stirling engine. The engine operates on a very different principle from that of the much better-known internal combustion engine. In the latter device, a chemical reaction takes place within a closed container, producing gases that drive a piston, whose motion can be used to perform useful mechanical work. In a Stirling engine, a gas is held within a sealed container, where no chemical change occurs. Instead, the gas is heated by some external source, such as a beam of solar energy, allowing the gas to perform work on a system. Stirling engines today have relatively few practical applications, one of which is in a number of solar energy systems.

Robert Stirling was born at Cloag Farm near Methven, Perthshire, Scotland, about 20 miles south of Glasgow. He was the third of eight children born to Patrick and Agnes Stirling, both from farming families from the area. He apparently became interested in science and engineering at a young age, perhaps because of his father's own involvement in the maintenance of threshing machines. Stirling decided, however, to pursue a career in the ministry and completed his studies in theology at the University of Edinburgh from 1805 to 1808. He then continued his studies in theology at the University of Glasgow, from which he graduated in 1816. He was then ordained a minister in the Church of Scotland and given an assistant post at the Laigh Kirk of Kilmarnock. In 1824, he was given his own parish at the Galston Church, where he remained until his retirement in 1878.

Throughout his life, Stirling maintained his interest in science and invention and made a number of contributions in the field of energy and optics. For example, during his years at Kilmarnock, he worked with the inventor Thomas Morton to design and make a number of optical instruments used in telescopes and for other purposes. He also continued his studies of heat engines. According to at least one observer, Pierre Gras (http://www.robertstirlingengine.com/history.php), Stirling's interest in this topic may have been motivated by the unhappy record of existing forms of the steam engine, which were inclined to be unstable and susceptible to explosions, often resulting in the death of workers and others in the vicinity of the event. Stirling's engine proved to be safer, more efficient, and quieter than existing steam engines.

For some time, a number of companies were interested in adopting the design used in the Stirling engine, and some invested substantial funds in bringing such engines to the marketplace. Such efforts were, however, almost universal failures as the vast majority of users preferred to stay with the earliest forms of steam engines that had made the Industrial Revolution both a possibility and an enormous success. Thus it is that even today Stirling engines tend to be used in only specific situations, the production of solar power being one of them. (A video illustrating the operation of a Stirling engine is available at https://www.youtube.com/watch?v=zCGTNArwJ0s.)

Stirling died at Galston, East Ayrshire, Scotland, on June 6, 1878. He was elected to the Scottish Engineering Hall of Fame in 2014.

U.S. Office of Energy Efficiency & Renewable Energy

The parentage of the Office of Energy Efficiency & Renewable Energy (EERE) dates to 1973, when President Richard Nixon declared his Project Independence, a program designed to make the United States less reliant on fossil fuels then purchased primarily from the oil-exporting states of the Middle East. He added a component to the existing Department of the Interior's program for research on coal, oil, and natural gas for similar studies of alternative fuels, although very little was actually done through that program.

In 1975, President Gerald Ford established the Energy Research and Development Administration (ERDA) as a way of placing greater emphasis on energy research issues, including renewable energy sources. Once again, relatively little was done on the development of alternative fuels. In 1977, President Jimmy Carter created the Department of Energy, into which was absorbed ERDA. In 2001, the agency was redesignated under its current name.

The EERE had a budget of $1.914 billion during fiscal year 2015, of which $778 million went to research on renewable fuels. The largest share of that money went to research on solar energy ($233 million), followed by research on bioenergy technologies ($225 million), wind ($107 million), hydrogen and fuel cells ($97 million), water ($61 million), and geothermal ($55 million). The major part of EERE's budget was spent on energy-efficient projects, such as research on more energy-efficient buildings, industrial operations, and vehicles.

EERE's activities are organized into about a half-dozen major categories: energy basics; states and local communities; technical assistance; energy analysis; jobs, education, and training; publications; and funding.

Energy basics is primarily a web-based service that provides general information on almost every conceivable topic in the area of energy efficiency and renewable energy, such as solar, wind, geothermal, and biomass sources of energy; industrial energy efficiency; passive solar building design; water heating; lighting and daylighting; and space heating and cooling.

States and local communities makes available information about energy efficiency and renewable energy programs and activities in all 50 U.S. states. It also describes a variety of federal and state programs on specific aspects of these topics, such as Advanced Manufacturing Office, Better Buildings Neighborhood Program, Building America, Building Energy Codes Program State Technical Assistance, Clean Cities, Clean Energy Application Centers, Database of State Incentives for Renewables and Efficiency, Geothermal Technologies Office, Green Power Network, Integrated Deployment State and Territory Projects, Standard Energy Efficiency Data Platform, Tribal Energy Program, State and Local Energy Efficiency Action Network, SunShot Initiative's Ask an Expert, Technical Assistance Program Solution Center, Weatherization & Intergovernmental Programs Office, and Wind Powering America.

Technical assistance offers individuals, governmental agencies, and businesses aid in dealing with a number of energy efficiency and renewable energy issues through agencies such as the State and Local Solution Center, Weatherization Assistance Program Technical Assistance Center, Tribal Energy Technical Assistance, FEMP Technical and Project Assistance Officer, Geothermal Maps, Geothermal Software and Data, Building America Solution Center, Building Energy Codes Program, Building Energy Software Tools Directory, Building Energy Codes Program State Technical Assistance, Fuel Cell Technologies Education, Combined Heat and Power Technical Assistance Partnerships, Technical Assistance for Manufacturers, Solar Outreach Partnership, Solar Technical Assistance Team, Alternative Fuels Data Center, Clean Cities Technical Assistance, Hydropower Resource Assessment and Characterization,

Marine and Hydrokinetic Resource Assessment and Characterization, and Wind Resource Maps and Data.

Energy analysis is an EERE website designed to help energy experts and policymakers get access to four major resources in decision making about efficiency and energy resource topics: data resources, market intelligence, energy systems analysis, and portfolio impacts analysis.

Jobs, education, and training is a resource for individuals wanting to learn more about education for and careers in energy efficiency and renewable energy resources.

Publications is a very large collection of books, articles, brochures, pamphlets, web pages, and other resources with information on energy efficiency and renewable energy.

Funding is one of EERE's most important functions since it has significant amounts of money available for the support of research and development of and demonstration projects on energy efficiency and renewable energy. The agency's website contains a page with detailed information about every aspect of the application and production features of funded projects.

EERE is not only a large and complex agency in and of itself, but it also has connections with a great many other federal and state organizations in the field of energy efficiency and renewable energy. Its web page called Offices (http://energy.gov/offices) is a good starting point for further research on these topics.

EERE's work in the field of solar energy is subdivided into seven major categories: photovoltaics, concentrating solar power (CSP), systems integration, reducing soft costs, bringing new technologies to market, solar energy resource center, and SunShot Incubator Program. For each of these areas, the agency provides basic information about the technology, examples of current research and development, funding opportunities, and other useful information related to the topic. The agency also provides a number of useful publications on many aspects of solar energy, such as the "2014 Concentrating Solar Power Report," a comprehensive review of the status of CSP in the United States; Solar Decathlon, a competition among colleges and university teams in the construction of solar houses; Financial Opportunities, an overview of grant and loan programs for a variety of solar projects; Guidance for Federal Agencies, a summary of laws, regulations, and other requirements that apply to all federal agencies; Solar Career Map, a listing and description of the types of careers available in solar energy; Active Solar Heating, a guide to and description of a form of solar technology useful in the construction of homes, offices, and other structures; and Solar Success Stories, concrete examples of the way in which solar energy has been put to use in the United States in specific circumstances.

The EERE library has an extensive list of resources available to the public at no cost, including audio, video, and image files; brochures; case studies; fact sheets; curricula; handbooks; letters; newsletters; and other type of publications. Examples of some of the materials available include "Capturing the Sun, Creating a Clean Energy Future" (brochure); "Best Practices Case Study: Grupe, Carsten Crossings, Rocklin CA" (fact sheet); "2011 Renewable Energy Data Book" (handbook); and "Solar Energy Technologies Program Newsletter."

Women in Solar Energy

Women in Solar Energy (WISE) was founded by Kristen Nicole, a longtime participant in many fields of the solar energy industry. Nicole has worked with utilities and regulatory agencies in the energy, water, telecommunications, and transportation sectors. Her areas of interest have included research on grid integration, development and energy policy, advanced power electronics technologies and applications, solar resource characterization, and impact of large-scale solar systems on transmission system planning and operations. During her career, she has been involved with the White House Council on Environmental Quality, the U.S. Department of Energy, Energy Information Administration, and the U.S. Peace Corps in Bangladesh. She holds a BA in international relations from Boston University and an MBA in accounting and strategic management from George Washington University.

In 2011 Nicole conceived of the notion of establishing an organization of women working in the solar industry. She was concerned by the changes that had occurred in the solar industry with which she had been associated for so many years. In particular, she saw a growing lack of diversity in the industry as more and more women began dropping out of the solar industry looking for careers that were more supportive of their needs and ambitions. A 2013 study, for example, showed that women made up less than 20 percent of the solar workforce, and minority groups were even less well represented with African Americans constituting less than 6 percent of workforce, Latinos about 15 percent, and Asian-Pacific Islands less than 7 percent. Nicole is now the executive director of WISE.

At its creation, WISE was defined as an organization with the goal of being "the networking center point of the solar energy industry, united towards a common goal of advancing women in all aspects of the solar energy industry and promoting diversity and forward thinking business practices in our community." The organization was founded on the belief that "the collective power of the female community is massive, and if we can all work together, the end result can be revolutionary."

WISE's activities are organized around six major themes: executive leadership, workforce, international, #nationWISE, Kidsun, and Go Solar. The executive leadership program is designed to promote the role of women in the solar industry and to advance their professional careers. Some activities included in this theme are professional networking events, scholarship opportunities for advanced education, working to include the number of women on solar boards, speaking opportunities and participation in webinars, and mentoring programs. The workforce theme is designed to advance the capabilities of and opportunities for women in the solar industry by providing mentoring programs, developing working relationships with all aspects of the industry, and providing scholarships for workers. The major component of the international theme is Solar Sister, a program designed to improve access to energy resources for women around the world by improving their economic opportunities.

The theme #nationWISE is a Twitter-based effort to recruit, retain, and retrain women for jobs in the solar industry. The Kidsun program is aimed at students at every educational level, from kindergarten through college. Its goal is to increase students' understanding of the nature of solar energy and its potential role in modern society. The program involves curriculum development on solar energy, scholarships for students interested in a solar career, mentoring programs, and speaking engagements. The Go Solar initiative is designed to promote a program known as Sungevity, whose goal it is to get individuals to convert their home energy system to a solar system. The program is available in 13 states and operates in cooperation with Lowe's Home Improvement stores.

Each year, WISE makes awards in a variety of solar-related fields, such as the Solar Educator of the Year (Kidsun project), Solar Developer of the Year (Workforce Development project), Solar Executive of the Year (Leadership Council), and Solar Humanitarian of the Year (International Development).

Hans K. Ziegler (1911–1999)

Ziegler was a German-born electrical engineer who was brought to the United States at the conclusion of World War II as part of a program known as *Operation Paperclip*. The purpose of that program was to get important German scientists with expertise that American research and industry could use and help them escape from postwar Germany before they were recruited by other nations, especially the Soviet Union. In the United States, he joined the U.S. Army Signal Corps at Fort Monmouth, New Jersey, where he worked in particular on propulsion and communication systems for space satellites. He was an early supporter of the use of nonconventional fuel sources for satellites and strongly recommended that they be employed in the launch of the first U.S. satellites in the late 1950s. Ziegler was at first frustrated in achieving this objective when the military contractor for the first satellites, the U.S. Navy, decided to make conventional chemical batteries the primary source of power for the first satellites. He eventually was successful, however, in convincing the navy to allow the use of solar batteries as backups for the chemical batteries. His faith in solar power was justified when the chemical batteries on the first satellites failed in a matter of weeks, while the solar batteries continued operating for many years. (Ziegler's story of his work with satellites and his campaign to make use of solar batteries is provided in detail in an article he wrote in 1981, "A Signal Corps Space Odyssey," available online at http://www.campevans.org/_CE/html/tac-ziegler.html.)

Hans K. Ziegler was born in Munich, Germany, on March 1, 1911. He attended the Technical College (now Technical University) of Munich, from which he received the technical equivalents of his BS, MS, and PhD degrees in 1932, 1934, and 1936, respectively, all in electrical engineering. While still a graduate student, he also served as assistant professor in electrical engineering at the Technical College. After completing his doctoral studies, Ziegler accepted an appointment as researcher at Rosenthal Isolatoren GmbH in Selb, Germany, a subsidiary of AEG (Allgemeine Elektrizitäts-Gesellschaft Aktiengesellschaft; German Electricity Corporation). Shortly after the beginning of World War II, Ziegler was made chief of the research and development department of the company, which worked on electrical components of bombs and other weapons.

Ziegler arrived in the United States in 1947 to take up his new assignment at the Signal Corps, where he continued to work for the next 30 years. During that time he held a number of positions, including scientific consultant, assistant director of research, director of the Astro-Electronics Division, and chief scientist of the Electronic Components Research Department of the Army Signal Corps Laboratory. In 1958 he was appointed director of the newly created Astro-Electronics Division and, a year later, chief scientist of the division. In 1963, he was appointed deputy for science and chief scientist of the U.S. Army Electronics Command and, in 1971, director of the U.S. Army Electronics Technology & Devices Laboratory, a post he held until his retirement in 1977.

Ziegler was a prolific researcher with many scholarly papers to his credit. He also served as a representative of the U.S. military at a number of national and international meetings and conferences. Ziegler was a delegate to the planning session of the International Geophysical Year in 1958 and a reviewer of research activities at the South Pole in 1964. He was awarded the Meritorious Civilian Service Award by the U.S. Department of Defense in 1963 and given the highest civilian award by the army upon his retirement. Ziegler died at Jersey Shore Medical Center, New Jersey, on December 11, 1999.

Introduction

Useful information about the status of solar energy in the United States and the rest of the world can often be gleaned from national, state, and local laws; court cases dealing with the topic; and statistics and data about the production and use of solar energy. This chapter provides some of this basic information on the topic of solar energy.

Data

The U.S. Energy Information Administration annually publishes its summary of energy activity in the United States for the previous year. Table 5.1 through Table 5.6 summarize basic information about the production and consumption of solar energy for years up to and including 2011, the most recent year for which data are available.

In 2010, at the request of the G8, a governmental forum of the world's leading economic powers, the International Energy Agency prepared a report about the prospects for solar power in the world's future energy equation to 2050. Table 5.7 summarizes the report's conclusion on that future.

As is the case with many personal and corporation activities in the United States, funding is available from many different branches of the federal government. Table 5.8 summarizes the

Solar panels, seen here on the side of a residential home, collect and concentrate sunlight, which is converted into electricity. (PRNewsFoto/Gaiam, Inc.)

kinds and sources of solar energy funding provided by the government as of 2013.

Concentrating solar power (CSP) is becoming an increasingly popular technology in many parts of the world. Table 5.9 summarizes CSP plants that have been opened for commercial use or are tied to a commercial electrical grid as of the end of 2010.

Table 5.1   Production of Solar Power in the United States, 1989–2011 (Quadrillion Btu).

Year	Solar Power	Coal[1]	Oil[1]
1989	0.055	21.360	16.117
1990	0.059	22.488	15.571
1991	0.062	21.636	15.701
1992	0.064	20.694	15.223
1993	0.066	20.336	14.494
1994	0.068	22.202	14.103
1995	0.069	22.130	13.887
1996	0.070	22.790	13.723
1997	0.070	23.310	13.658
1998	0.069	24.045	13.236
1999	0.068	23.295	12.451
2000	0.066	22.735	12.358
2001	0.064	23.547	12.282
2002	0.063	22.732	12.163
2003	0.062	22.094	12.026
2004	0.063	22.852	11.503
2005	0.063	23.185	10.963
2006	0.068	23.790	10.801
2007	0.076	23.493	10.721
2008	0.089	23.851	10.509
2009	0.098	21.624	11.348
2010	0.126	22.038	11.593
2011	0.158	22.181	11.986

[1]For purposes of comparison.
Source: Table 1.2. Primary Energy Production by Source, Selected Years, 1949–2011. Annual Energy Review 2011. U.S. Energy Information Administration.http://www.eia.gov/totalenergy/data/annual/pdf/aer.pdf. Accessed on February 3, 2015.

Table 5.2 Solar Energy Consumption, Residential, 1989–2011 (Trillion Btu).

Year	Solar	Natural Gas[1]	Oil[1]
1989	52	4,899	1,660
1990	56	4,491	1,394
1991	57	4,667	1,381
1992	60	4,805	1,414
1993	61	5,063	1,439
1994	63	4,960	1,408
1995	64	4,954	1,374
1996	65	5,354	1,484
1997	64	5,093	1,422
1998	64	4,646	1,304
1999	63	4,835	1,465
2000	61	5,105	1,554
2001	59	4,889	1,529
2002	57	4,995	1,457
2003	57	5,209	1,519
2004	57	4,981	1,520
2005	58	4,946	1,451
2006	63	4,476	1,224
2007	70	4,835	1,254
2008	80	5,010	1,243
2009	89	4,883	1,176
2010	114	4,883	1,142
2011	140	4,830	1,139

[1]For purposes of comparison.
Source: Table 2.1b. Residential Sector Energy Consumption Estimates, Selected Years, 1949–2011. Annual Energy Review 2011. U.S. Energy Information Administration. http://www.eia.gov/totalenergy/data/annual/pdf/aer.pdf. Accessed on February 3, 2015.

Table 5.3　Electric Power Consumption, 1989–2011 (Trillion Btu).

Year	Solar	Coal[1]	Oil[1]
1989	3	16,137	3,173
1990	4	16,261	3,309
1991	5	16,250	3,377
1992	4	16,466	3,512
1993	5	17,196	3,538
1994	5	17,261	3,977
1995	5	17,466	4,302
1996	5	18,429	3,862
1997	5	18,905	4,126
1998	5	19,216	4,675
1999	5	19,279	4,902
2000	5	20,220	5,293
2001	6	19,614	5,458
2002	6	19,783	5,767
2003	5	20,185	5,246
2004	6	20,305	5,595
2005	6	20,737	6,015
2006	5	20,462	6,375
2007	6	20,808	7,005
2008	9	20,513	6,829
2009	9	18,225	7,022
2010	12	19,133	7,527
2011	18	17,986	7,740

[1]For purposes of comparison.
Source: Table 2.1f. Electric Power Sector Energy Consumption, Selected Years, 1949–2011. Annual Energy Review 2011. U.S. Energy Information Administration. http://www.eia.gov/totalenergy/data/annual/pdf/aer.pdf. Accessed on February 3, 2015.

Table 5.4   Total Solar Energy Production in the United States, 1989–2011
(Trillion Btu).

Year	Solar	Total Renewable	Percentage of Renewable[1]
1989	55	6,235	0.9
1990	59	6,041	1.0
1991	62	6,069	1.0
1992	64	5,821	1.1
1993	66	6,083	1.1
1994	68	5,988	1.1
1995	69	6,558	1.0
1996	70	7,012	1.0
1997	70	7,018	1.0
1998	69	6,494	1.1
1999	68	6,517	1.0
2000	66	6,104	1.1
2001	64	5,164	1.2
2002	63	5,734	1.1
2003	62	5,982	1.0
2004	63	6,070	1.0
2005	63	6,229	1.0
2006	68	6,599	1.0
2007	76	6,509	1.2
2008	89	7,202	1.2
2009	98	7,616	1.3
2010	126	8,136	1.5
2011	158	9,236	1.7

[1]Calculated from table data.
Source: Table 10.1. Renewable Energy Production and Consumption by Primary Energy Source, Selected Years, 1949–2011. Annual Energy Review 2011. U.S. Energy Information Administration. http://www.eia.gov/totalenergy/data/annual/pdf/aer.pdf. Accessed on February 3, 2015.

Table 5.5   End Use for Solar Thermal Collector Shipments, 2001–2009
(Thousand Square Feet).

Year	Pool Heating	Water Heating	Space Heating	Space Cooling	Combined Heating	Process Heating	Electrical Generation
2001	10,797	274	70	0	12	34	2
2002	11,073	423	146	(s)	17	4	0
2003	10,800	511	76	(s)	23	34	0
2004	13,634	452	13	0	16	0	0
2005	15,041	640	228	2	16	0	114
2006	15,362	1,136	330	3	66	0	3,847
2007	2,076	1,393	189	13	73	27	6
2008	11,973	1,978	186	18	148	50	361
2009	8,934	1,992	150	10	137	608	389

Date prior to 2006 for domestic and export use; after 2006, for domestic only.

(s) = Less than 0.5 thousand square feet

Source: Table 10.7. Solar Thermal Collector Shipments by Market Sector, End Use, and Type, 2001–2009. Annual Energy Review 2011. U.S. Energy Information Administration. http://www.eia.gov/totalenergy/data/annual/pdf/aer.pdf. Accessed on February 3, 2015.

Table 5.6   Average Price of Photovoltaic Cells and Modules, 2003–2012.

Year	Price per Cell	Price per Module
2003	1.86	3.17
2004	1.92	2.99
2005	2.17	3.19
2006	2.03	3.50
2007	2.22	3.37
2008	1.94	3.49
2009	1.27	2.79
2010	1.13	1.96
2011	0.92	1.59
2012	1.00	1.15

Source: Table 4. Solar Photovoltaic Cell/Module Shipments Report. U.S. Energy Information Administration. http://www.eia.gov/renewable/annual/solar_photo/. Accessed on February 15, 2015.

Table 5.7   Future Scenario for Solar Power as Total of World Electrical
Consumption (%)

	2020	2030	2020	2040	2050
United States	0.1	6.8	0.1	15	18
Other OECD Americas	0.2	1.5	0.2	5.8	8.5
European Union	0.1	1.6	0.1	3.0	3.7
Other OECD	0.1	0.7	0.1	1.9	2.4
China	0.1	1.3	0.1	3.5	5.4
India	0.8	6.0	0.8	15	21
Africa	0.5	8.6	0.5	22	26
Middle East	0.5	14	0.5	29	40
Other Developing Asia	0.0	0.0	0.0	0.4	1.1
Eastern Europe and FSU	0.0	0.0	0.0	0.0	0.0
Non-OECD Americas	0.2	0.5	0.2	1.6	2.6

Source: Data from **OECD/IEA. ©OECD/IEA.** IEA Publishing. License: www.iea
.org/t&c/termsandconditions. Used by permission.

Table 5.8   Funding Opportunities for Solar Energy from the Federal
Government (as of 2013).

Agency	Program	Qualifying Technologies
Department of Energy Office of Energy Efficiency and Renewable Energy	Solar Energy Technologies Program	Solar
	Building Technologies Program	Passive solar Photovoltaics
	Energy Efficiency and Renewable Energy Technology Deployment, Demonstration, and Commercialization Grant Program	Solar
	Renewable Energy Production Initiative	Solar thermal electric Photovoltaics
	Renewable Energy and Research Program	Solar
	Tribal Energy Program	Passive solar space heat Solar water heat Solar space heat Photovoltaics
	Federal Energy Management Program	Solar

(Continued)

Table 5.8 (Continued)

Agency	Program	Qualifying Technologies
	Loan Guarantee Program	Solar thermal electric Solar thermal process heat Photovoltaics
Department of the Treasury	Residential Renewal Energy Tax Credit	Solar water heat Photovoltaics
	Business Energy Investment Tax Credit	Solar water heat Solar space heat Solar thermal electric Solar thermal process heat Photovoltaics Solar hybrid lighting
	Qualifying Advanced Engineering Manufacturing Investment Tax Credit	Solar water heat Solar thermal electric Photovoltaics
	Residential Energy Conservation Subsidy Exclusion (Corporate)	Solar water heat Solar space heat Photovoltaics
	Residential Energy Conservation Subsidy Exclusion (Personal)	Solar water heat Solar space heat Photovoltaics
	Qualified Energy Conservation Bonds	Solar thermal electric Photovoltaics
	Modified Accelerated Cost-Recovery Systems	Solar water heat Solar space heat Solar thermal electric Solar thermal process heat Photovoltaics Solar hybrid lighting
Department of Agriculture	Rural Energy for America Program	Solar water heat Solar space heat Solar thermal electric Photovoltaics
Department of Housing and Urban Development	Energy Efficient Mortgages	Passive solar space heat Solar water heat Solar space heat Photovoltaics
Department of Veterans Affairs	Energy Efficient Mortgages	Passive solar space heat Solar water heat Solar space heat Photovoltaics

Source: Cunningham, Lynn J., and Beth A. Roberts. 2013. "Renewable Energy and Energy Efficiency Incentives: A Summary of Federal Programs." Congressional Research Service. http://fas.org/sgp/crs/misc/R40913.pdf. Accessed on March 21, 2015.

Table 5.9   Commercial CSP Plants Installed Worldwide, as of December 2010.

Name	Location	Type	Year Installed	Capacity (MW)
SEGS I—IX	California	Trough	1985–1990	354
Inditep	Spain	Trough	2005	1.2*
APS Saguaro	Arizona	Trough	2006	1
Nevada Solar One	Nevada	Trough	2007	64
PS10	Spain	Tower	2007	11
Kimberlina	California	CLFR	2008	7*
Andasol 1	Spain	Trough	2008	50
Liddell	Australia	CLFR	2008	3
Sierra Sun Tower	California	Tower	2009	5*
Holaniku	Hawaii	Trough	2009	2*
Stadtwerke Julich	Germany	Tower	2009	1.5*
Puerto Errado 1	Spain	Linear Fresnel	2009	1.4*
Puertollano Ibersol	Spain	Trough	2009	50
La Risca	Spain	Trough	2009	50
PS20	Spain	Tower	2009	20
Holaniku	Hawaii	Trough	2009	2
Maricopa Solar	Arizona	Dish	2010	1.5*
IEECAS Badaling	China	Tower	2010	1.5*
Cameo	Colorado	Trough	2010	1*
Casa del Ángel Termosolar	Spain	Stirling	2010	1*
Himin Yanqing	China	Tower	2010	1*
Martin	Florida	Trough	2010	75
Andasol 2	Spain	Trough	2010	50
Extresol 1	Spain	Trough	2010	50
Solnova 1	Spain	Trough	2010	50
Solnova 3	Spain	Trough	2010	50
Solnova 4	Spain	Trough	2010	50

(Continued)

Table 5.9 (Continued)

Name	Location	Type	Year Installed	Capacity (MW)
La Florida	Spain	Trough	2010	50
Majadas	Spain	Trough	2010	50
La Dehesa	Spain	Trough	2010	50
Palma Del Rio II	Spain	Trough	2010	50
Extresol-2	Spain	Trough	2010	50
Manchasol-1	Spain	Trough	2010	50
Ain Beni Mathar	Morocco	Trough	2010	20
Al Kuraymat	Egypt	Trough	2010	20
Archimede	Italy	Trough	2010	5

*Connected to grid

CLFR = Compact linear Fresnel reflector

Source: SunShot Vision Study. February 2012. National Renewable Energy Laboratory. http://energy.gov/sites/prod/files/2014/01/f7/47927_chapter2.pdf. Accessed on March 20, 2015.

Documents

The Heliocaminus and Solar Rights (529–534)

In both ancient Greece and Imperial Rome, many citizens had become accustomed to building their homes in such a way as to capture the Sun's rays in the winter, and using solar radiation to heat their homes, and to deflect those rays in the summer, keeping the house cool. Over time, questions began to arise about structures that were built adjacent to such solar passive homes, called heliocaminum (singular) or heliocamina (plural). If the newer structures blocked solar radiation from reaching the older homes, of course, the whole point of a heliocaminum was defeated. As early as the sixth century, laws were adopted to prevent the placement of objects that would prevent sunlight from reaching solar passive homes. Probably the most important of these laws can be found in the Justinian code, promulgated between 529 and 534 CE.

If a man plants a tree so as to shut out [a dominant owner's] light, it may perfectly well be said that he acts in violation of the servitude created; even a tree causes a smaller extent of sky to be seen. But if what is placed on the spot in no respect impedes the light, but only keeps off the sunshine, then, if it is done in a place where it was more agreeable to be without sunshine, it may be said that the party does not do anything in violation of the servitude ; but if he does it over against a room meant to be exposed to the sun (heliocaminus), or a sundial, then it must be said that, by causing shade in a place where sunshine was indispensable, he is acting in violation of the servitude created. 1. If a man, on the other hand, should remove a building or, say, branches of trees, with the result that a spot which up to that time was in the shade comes to be fully exposed to the sunshine, he does not violate the servitude; the servitude to which he was subject was to the effect that he should not obscure the light, but, in the case in question, he does not obscure the light, he lets light in to an excessive extent. 2. However, there are cases in which it may be said that even a man who removes or lowers a building does obscure the light, where, for example, light found its way into the house by reflexion or some sort of repercussion.

Source: *The Digest of Justinian.* Charles Henry Munro, translator. Cambridge: University Press, 1909, 72.

Solar Energy Research, Development, and Demonstration Act of 1974

In the wake of the Arab oil embargo of 1973, the U.S. Congress took a number of critical steps in the following year that marked a critical turning point in the history of energy policy in the United States. The Congress passed the Energy Reorganization Act of 1974 that created the Energy Research and Development Administration (ERDA), the immediate predecessor of the U.S. Department of Energy (DOE), as well as a group of four other energy-related bills, one of which was the Solar Energy Research, Development, and Demonstration Act of 1974. That act, for the first time in history, laid out a clearly defined and specific agenda for the role of solar energy in the nation's future. The essential sections of that act were as follows:

SEC. 4. (a) There is hereby established the Solar Energy Co-ordination and Management Project. . . .

(2) The President shall designate one member of the Project to serve as Chairman of the Project. . . .

Resource Determination and Assessment

SEC. 5. (a) The Chairman shall initiate a solar energy resource determination and assessment program with the objective of making a regional and national appraisal of all solar energy resources, including data on insolation, wind, sea thermal gradients, and potentials for photosynthetic conversion. The program shall emphasize identification of promising areas for commercial exploitation and development.

The specific goals shall include—

[Seven specific goals are then listed, examples of which are the following:]

(1) the development of better methods for predicting the availability of all solar energy resources, over long time periods and by geographic location;

(2) the development of advanced meteorological, oceanographic, and other instruments, methodology, and procedures necessary to measure the quality and quantity of all solar resources on periodic bases; . . .

Research and Development

SEC. 6. (a) The Chairman shall initiate a research and development program for the purpose of resolving the major technical problems inhibiting commercial utilization of solar energy in the United States.

(b) In connection with or as a part of such program, the Chairman shall—

   (1) conduct, encourage, and promote scientific research and studies to develop effective and economical processes and equipment for the purpose of utilizing solar energy in an acceptable manner for beneficial uses;

   (2) carry out systems, economic, social, and environmental studies to provide a basis for research, development and demonstration planning and phasing; and

   (3) perform or cause to be performed technology assessments relevant to the utilization of solar energy.

(c) The specific solar energy technologies to be addressed or dealt with in the program shall include—

   (1) direct solar heat as a source for industrial processes, including the utilization of low-level heat for process and other industrial purposes;

   (2) thermal energy conversion, and other methods, for the generation of electricity and the production of chemical fuels;

   (3) the conversion of cellulose and other organic materials (including wastes) to useful energy or fuels;

   (4) photovoltaic and other direct conversion processes;

   (5) sea thermal gradient conversion;

   (6) windpower conversion;

   (7) solar heating and cooling of housing and of commercial and public buildings; and

   (8) energy storage.

Demonstration

SEC. 7. (a) The Chairman is authorized to initiate a program to design and construct, in specific solar energy technologies (including, but not limited to, those listed in section (6) (c), facilities or powerplants of sufficient size to demonstrate the technical and economic feasibility of utilizing the various forms of solar energy. The specific goals of such programs shall include—

(1) production of electricity from a number of power-plants, on the order of one to ten megawatts each;

(2) production of synthetic fuels in commercial quantities;

(3) large-scale utilization of solar energy in the form of direct heat;

(4) utilization of thermal and all other byproducts of the solar facilities;

(5) design and development of hybrid systems involving the concomitant utilization of solar and other energy sources; and

(6) the continuous operation of such plants and facilities for a period of time.

[The act also established the Solar Energy Research Institute, which was, in 1991, renamed the National Energy Research Institute, with a vastly expanded mission in all areas of renewable energy.]

Solar Energy Research Institute

SEC. 10. (a) There is established a Solar Energy Research Institute, which shall perform such research, development, and related functions as the Chairman may determine to be necessary or appropriate in connection with the Project's activities under this Act or to be otherwise in furtherance of the purpose and objectives of this Act.

Source: Public Law 93–473.

Public Utility Regulatory Policies Act of 1978 (P.L. 95-617)

An important event in the history of solar power promotion in the United States was the passage of the Public Utility Regulatory Policies Act of 1978. That act for the first time required that large public utilities be equipped to accept excess electricity produced by other sources, such as small electrical companies or individual residential owners of renewable energy resources, such as rooftop solar facilities. This legislation significantly increased the willingness of individuals and small companies to produce electricity with their own solar facilities, a factor that continues to encourage the growth of solar energy production in the United States.

Sec. 202. Interconnection.

Part II of the Federal Power Act is amended by adding the following new section at the end thereof:

"Certain Interconnection Authority"

"SEC. 210. (a)(1) Upon application of any electric utility, Federal power marketing agency, qualifying cogenerator, or qualifying small power producer, the Commission may issue an order requiring—

"(A) the physical connection of any cogeneration facility, any small power production facility, or the transmission facilities of any electric utility, with the facilities of such applicant,

"(B) such action as may be necessary to make effective any physical connection described in subparagraph (A), which physical connection is ineffective for any reason, such as inadequate size, poor maintenance, or physical unreliability.

"(C) such sale or exchange of electric energy or other coordination, as may be necessary to carry out the purposes of any order under subparagraph (A) or (B), or

"(D) such increase in transmission capacity as may be necessary to carry out the purposes of any order under subparagraph (A) or (B).

Source: Public Law 95–617.

Energy Tax Act of 1978 (P.L. 95-618)

A key piece of legislation that has made possible the continued growth of solar and other forms of renewable energy in the United States was the Energy Tax Act of 1978. That act included a number of provisions designed to wean the United States away from its heavy dependence on foreign supplies of fossil fuels. One of the key elements of the act was the creation of so-called tax incentives, financial provisions designed to encourage the purchase and use of renewable energy resources. The following section of the act creates these incentives, which have been retained or renewed in one form or another almost continuously since the passage of the 1978 law. (The references to "amending" refer to changes in the tax code provided for by this law.)

Title I—Residential Energy Credit
Sec. 101. Residential Energy Credit.

(a) GENERAL RULE.—Subpart A of part I V of subchapter A of chapter 1 (relating to credits allowable) is amended by inserting after section 44B the following new section:

"Sec. 44c. Residential Energy Credit."

"(a) GENERAL RULE.—In the case of an individual, there shall be allowed as a credit against the tax imposed by this chapter for the taxable year an amount equal to the sum of—

"(1) the qualified energy conservation expenditures, plus

"(2) the qualified renewable energy source expenditures.

"(b) QUALIFIED EXPENDITURES.—For purposes of subsection (a)—

"(1) ENERGY CONSERVATION.—In the case of any dwelling unit, the qualified energy conservation expenditures are 15 percent of so much of the energy conservation expenditures made by the taxpayer during the taxable year with respect to such unit as does not exceed $2,000.

"(2) RENEWABLE ENERGY SOURCE.—In the case of any dwelling unit, the qualified renewable energy source expenditures are the following percentages of the renewable energy source expenditures made by the taxpayer during the taxable year with respect to such unit:

"(A) 30 percent of so much of such expenditures as does not exceed $2,000, plus

"(B) 20 percent of so much of such expenditures as exceeds $2,000 but does not exceed $10,000.

[Following parts of this section deal with the details of calculating the tax credit.]

Source: Public Law 95–618.

California Solar Power Law (1978)

The value of solar power in meeting societal needs for energy dates to the last quarter of the twentieth century in some states of the United States. As early as the 1970s, for example, the state of California was beginning to consider legal provisions that might be needed for the protection and promotion of various types of renewable energy systems in the state. An example of the legislation adopted during this period is the Solar Rights Act of 1978, which codified a number of aspects of the research, development, installation, and use of solar power in the state. An important feature of the act was a provision that prevented homeowners' associations and other groups and individuals from adopting regulations that would prevent individuals from installing their own solar power systems. A provision of the 1978 law that exempted cities, counties, and other municipalities from provisions of the act was revoked in 2003.

714. (a) Any covenant, restriction, or condition contained in any deed, contract, security instrument, or other instrument affecting the transfer or sale of, or any interest in, real property, and any provision of a governing document, as defined in Section 4150 or 6552, that effectively prohibits or restricts the installation or use of a solar energy system is void and unenforceable.

(b) This section does not apply to provisions that impose reasonable restrictions on solar energy systems. However, it is the policy of the state to promote and encourage the use of solar energy systems and to remove obstacles thereto. Accordingly, reasonable restrictions on a solar energy system are those restrictions that do not significantly increase the cost of the system or significantly decrease its efficiency or specified performance, or that allow for an alternative system of comparable cost, efficiency, and energy conservation benefits.

(c) (1) A solar energy system shall meet applicable health and safety standards and requirements imposed by state and local permitting authorities, consistent with Section 65850.5 of the Government Code. . . .

The above provision was extended to municipalities in a 2005 action:

17959.1. . . .

(b) A city or county may not deny an application for a use permit to install a solar energy system unless it makes written findings based upon substantial evidence in the record that the proposed installation would have a specific, adverse impact upon the public health or safety, and there is no feasible method to satisfactorily mitigate or avoid the specific, adverse impact. This finding shall include the basis for the rejection of potential feasible alternatives of preventing the adverse impact.

Sources: Civil Code Section 707-714.5 and Health and Safety Code Section 17950-17959.6.

New Hampshire Solar Easement Law (1985)

Individual states have been at the forefront of solar law and policies, often preceding federal action by many years. One such early law is the solar easement law adopted by the State of New Hampshire in 1985, in which property owners are permitted to obtain easements for access to solar radiation to operate their solar facilities, as they are able to do for other types of easements. The law reads as follows:

I. A solar skyspace easement may be acquired and transferred and shall be recorded in t the same manner as any other conveyance of an interest in real property. The easement shall run with the land benefited and burdened and shall terminate upon the conditions stated in the instrument creating the easement or upon court decree based upon abandonment or changed conditions or as provided in RSA 477:26; provided, however, that no planning board may require a landowner to grant an easement.

II. An instrument creating a solar skyspace easement shall include, but not be limited to:

(a) A description of the vertical and horizontal angles, expressed in degrees and measured from the site of the solar energy system, at which the solar skyspace easement extends over the real property subject to the solar skyspace easement, or any other description which describes the 3-dimensional space, or the place and times of day in which an obstruction to solar energy is prohibited or limited;

(b) Terms or conditions under which the easement is granted or shall be terminated;

(c) Provisions for compensation of the benefited landowner in the event of interference with the enjoyment of the easement or compensation of the burdened landowner for maintaining the easement; and

(d) A description of the real property subject to the solar skyspace easement and a description of the real property benefiting from the solar skyspace easement.

III. A solar skyspace easement shall not terminate within 10 years after its creation unless an earlier termination is expressly stated in the instrument or is otherwise negotiated by the owners of the benefited and burdened land. The easement may be enforced by proceedings in equity and by actions at law for damages.

Source: New Hampshire Statutes. Chapter 477, Section 477:50.

Sher v. Leiderman (1986)

One of the most persistent issues relating to the development and use of solar power systems involves access to sunlight with which to operate such systems. The usual situation is one in which a person, a family, or a small business decides to construct a solar power system on top of their home, in their backyard, or at some other location on their property. Over time, however, changes occur on adjacent properties that restrict the amount of sunlight that can flow to these systems and the system owners then sue to have the offending obstruction removed. The obstruction may take any number of forms, most commonly the construction of a tall new building or the planting or growth of large trees. In English common law, such interference with a person's access to sunlight was restricted by a doctrine known as Ancient Lights, which said that all people were entitled to have access to reasonable amounts of sunlight. That doctrine has, for the most part, now been rejected by courts in the United States. In the case cited here, two couples bought adjacent plots of land in 1962, after which one couple (the Shers) built *a home that made extensive use of solar energy, and the Leidermans planted trees, which eventually grew large enough to shade the Shers' solar system. The Shers sued to have the Leidermans cut down or, at least, trim their trees. As had most courts at the time of this decision, the California court that heard this case ruled against the Shers, saying that they had no legal right to the sunlight they needed to operate their solar system (omitted citations are indicated with triple asterisks, ***).*

[The Shers] appeal presents an issue of first impression in this state, namely, whether an owner of a residence designed to make use of solar energy can state a cause of action for private nuisance when trees on his neighbor's property interfere with his solar access. We determine that California nuisance law does not provide a remedy for blockage of sunlight, and, for reasons discussed below, we decline to expand existing law. . . .

The court opened its discussion by observing that the interests protected by nuisance law are broadly defined to include practically any disturbance of the enjoyment of property. . . .

Although obstruction of sunlight would appear to fall within this concept of a private nuisance as "a nontrespassory invasion of another's interest in the use and enjoyment of land" ***, courts have traditionally refused to consider a landowner's access to sunlight a protected interest. This judicial posture stems, in the Wisconsin *[supreme]* court's view, from the early repudiation by American courts of the English common law "Doctrine of Ancient Lights," under which a landowner could acquire a prescriptive easement to receive sunlight over adjoining property. Such a doctrine was ill-suited to conditions existing in the early part of this century in a new and rapidly growing country. At that time society had a significant interest in encouraging unrestricted land development. Moreover a landowner's rights to use his land were virtually unlimited; it was thought that he owned to the center of the earth and up to the heavens. In contrast, light had little social importance beyond its value for aesthetic enjoyment or illumination. . . .

Because of this inversion of social priorities over the years, it is urged that interference with solar access should no longer be considered a "mere" obstruction of light, as it once was; today it may in fact amount to substantial and perceptible harm, certainly no less substantial than the harm caused by other recognized nuisances such as unpleasant odors, noise, smoke, vibrations or dust.

[The court explains that the state of California has attempted to deal with this problem by passing the California Solar Shade Control Act of 1978.]

The California Legislature has already seen fit to carve out an exception to established nuisance law, in the form of the California Solar Shade Control Act. *** Under this law, after January 1, 1979, a property owner *** cannot plant a tree, or allow one to grow, which will shade more than 10 percent of a neighbor's solar collector between the hours of 10 a.m. and 2 p.m. The property owner must be given 30 days notice, after which, if he has failed to remove or alter the offending tree(s), he is guilty of a public nuisance, punishable by a fine not to exceed $1,000 per day for each day the violation continues.

[To the court, this act provides the precise limitations involved in obstructing access to sunlight for solar power system. It concludes that:]

We are unwilling to intrude into the precise area of the law where legislative action is being taken. If the Legislature intended to limit its protection of solar access to those situations circumscribed by the Solar Shade Control Act, our expansion of the nuisance law beyond those bounds would be unwarranted. On the other hand, the Solar Shade Control Act may well represent the initial phase of a more comprehensive legislative plan to guarantee solar access; in that case, judicial interference could undermine the orderly development of such a scheme.

[On that basis, and considering all other relevant facts, the court rules in favor of the Leidermans.]

Source: *Sher v. Leiderman*, 181 Cal. App. 3d 867 (1986)

Palos Verdes Homes Association v. Rodman, 182 Cal. App. 3d 324, 324–329 (1986)

One of the ongoing legal issues associated with the use of solar energy has to do with the rights of individual property owners to install solar devices (usually solar panels) on their own property, with or without the permission of governmental or nongovernmental entities to which they may be answerable, such as city, town, county, or state governments or homeowners' associations. One of the landmark decisions in this field came as the result of a 1986 court case in which a homeowner, Stacy Rodman, decided to install a passive solar water heating system on the roof of his home, although the Covenants, Conditions, and Regulations (CC&Rs) of the homeowners' association to which he belonged provided specific instructions for the installation of such systems, which Rodman's system did not appear to meet. When the association ordered Rodman to remove his solar system, he sued the association, claiming that the CC&Rs were not reasonable and in violation of California law and policy that encouraged the use of solar systems in the state. Both the lower court and the appeals court in which the case was heard ruled in favor of the homeowners' association on the basis that its CC&Rs imposed "reasonable" limitations on the type of solar system that could be installed in the neighborhood. [Asterisks indicate the omission of text or reference materials.]

Rodman and Servamatic (both as appellants) have joined in the appeal from this judgment *[of the lower court]*. They urge us to find that as a matter of law the Association's solar unit guidelines are in violation of the spirit and intent of section 714 *[of the California Civil Code]*. Specifically, appellants argue that these guidelines "effectively prohibit or restrict" installation of solar energy units in that they significantly increase the cost of a system, decrease a system's efficiency and do not allow for "an alternative system of comparable cost and efficiency," all in contravention of the policy stated in section 714, " . . . to promote and encourage the use of solar energy systems and to remove obstacles. . . . "

* * *

The Association counters that the pertinent and controlling language of section 714 is that which finds "reasonable restrictions" to include those ". . . which allow for an alternative system of comparable cost and efficiency." (1a) The Association argues that the solar unit guidelines do not prohibit all solar units but are formulated to promote the installation of solar units which are comparable in costs and aesthetically acceptable. We concur.

(2) The right to enforce covenants that require approval of construction has long been recognized in California. *** (3) The issue here is whether the Association's Guidelines are a "reasonable restriction" on the installation of solar units, as required by section 714. This is a question of fact to be determined by the trier of fact. Its conclusion will not be disturbed unless unsupported by substantial evidence. ***

(1b) The evidence presented at the court trial included testimony by William Nelson Rowley, Ph.D., in mechanical engineering. Dr. Rowley, an engineer for 30 years and member of numerous professional organizations, has been designing and selling solar systems since 1974, designing and/or installing more than 250 systems. In October 1983, he was hired as a consultant to the Association to study and determine whether the Association's solar guidelines complied with section 714. His study included a comparison of the costs of various solar systems, including appellant's; their positive and/or negative aspects; capacity; weight; insulation; tank temperature design; efficiency; output and warranties. Based on his study comparing 26 systems installed on the Palos Verdes Peninsula with appellant's ICS *[integral collector system]*, he concluded the solar units permitted by the Association guidelines were comparable to the ICS in performance and costs.

Given this evidence and the fact that appellant's own witness agreed that the various solar systems discussed by Dr. Rowley were comparable, we cannot say as a matter of law that the trial court erred. The evidence before the court, partially summarized above, clearly supports the judgment rendered.

The judgment is affirmed.

Source: *Palos Verdes Homes Assn. v. Rodman*, 182 Cal. App. 3d 324 (Cal. Ct. App. 1986)

Energy Policy Act of 1992 (P.L. 102-486)

The first tax incentives for the use of renewable energy approved by the U.S. Congress were those given for investments made in the equipment and installation used in producing wind, solar, geothermal, and other types of renewable energy. In 1992, the Congress adopted a somewhat different type of tax incentive policy, one based on the amount of energy produced by renewable energy resources, rather than on the investment in such energy resources. Solar energy was specifically excluded from the 1992 act, although it was later added to the production tax credit (PTC) program more than a decade later (see American Jobs Creation Act of 2004). The principles behind the PTC, then, are as follows. (Amendments referred to in this selection are with regard to the Federal Power Act of 1935.)

Sec. 1914. Credit for Electricity Produced from Certain Renewable Sources.

(a) IN GENERAL.—Subpart D of part IV of subchapter A of chapter 1 is amended by adding at the end thereof the following new section:

Sec. 46. Electricity Produced from Certain Renewable Resources.

"(a) GENERAL RULE.-For purposes of section 38, the renewable electricity production credit for any taxable year is an amount equal to the product of-

"(1) 1.5 cents, multiplied by

"(2) the kilowatt hours of electricity-

"(A) produced by the taxpayer-

"(i) from qualified energy resources, and

"(ii) at a qualified facility during the 10-year period beginning on the date the facility was originally placed in service, and

"(B) sold by the taxpayer to an unrelated person during the taxable year. . . .

[Succeeding sections deal with the details involved in calculating the actual tax credit.]

"(c) DEFINITIONS.—For purposes of this section-

"(1) QUALIFIED ENERGY RESOURCES.—The term 'qualified energy resources' means-

"(A) wind, and

"(B) closed-loop biomass.

Source: Public Law 102–486.

American Jobs Creation Act of 2004 (P.L. 108–357)

Solar energy had been excluded from provisions of the production tax credit (PTC) by the Energy Policy Act of 1992. That omission was remedied in 2004 when Congress passed the American Jobs Creation Act. The act also provided a "sunset" provision of January 1, 2006, meaning that this tax credit expired on that date. In such cases, the Congress often renews the tax credit in time for it to be continued after its original deadline, a step that Congress did not take in this case. (See Energy Policy Act of 2005.)

Sec. 710. Expansion of Credit for Electricity Produced from
Certain Renewable Resources.

(a) EXPANSION OF QUALIFIED ENERGY RESOURCES.—Subsection (c) of section 45 (relating to electricity produced from certain renewable resources) is amended to read as follows:

*"(c) QUALIFIED ENERGY RESOURCES AND RE-FINED COAL.—For purposes of this section:

"(1) IN GENERAL.—The term 'qualified energy resources' means—

"(A) wind,

"(B) closed-loop biomass,

"(C) open-loop biomass,

"(D) geothermal energy,

"(E) solar energy,

"(F) small irrigation power, and

"(G) municipal solid waste. . . .

[The definition of each term then follows.]

"(4) GEOTHERMAL OR SOLAR ENERGY FACIL-ITY.—In the case of a facility using geothermal or solar energy to produce electricity, the term 'qualified facility' means any facility owned by the taxpayer which is originally placed in service after the date of the enactment of this paragraph and before January 1, 2006. Such term shall not include any property described in section 48(a)(3) the basis of which is taken into account by the taxpayer for purposes of determining the energy credit under section 48.

No section (b) occurs at this point.

Source: Public Law 108–357.

Energy Policy Act of 2005 (P.L. 109-58) (2005)

The key piece of federal legislation dealing with solar power (and all other forms of renewable energy) is the Energy Policy Act of 2005. The act establishes federal policy for the development and use of renewable forms of energy and provides a system of investment tax credits (ITCs) for the purpose of promoting the installation and use of commercial and residential solar power systems. The act was later amended and updated by the Tax Relief and Health Care Act of 2006 (P.L. 109-432), the Emergency Economic Stabilization Act of 2008 (P.L. 110-343), and the American Recovery and Reinvestment Act of 2009 (P.L. 111-5). Some relevant parts of the original 2005 act are excerpted here.

§ 3177. Use of Photovoltaic Energy in Public Buildings

(a) PHOTOVOLTAIC ENERGY COMMERCIALIZATION PROGRAM.—

(1) IN GENERAL.—The Administrator of General Services may establish a photovoltaic energy commercialization program for the procurement and installation of photovoltaic solar electric systems for electric production in new and existing public buildings.

(2) PURPOSES.—The purposes of the program shall be to accomplish the following:

(A) To accelerate the growth of a commercially viable photovoltaic industry to make this energy system available to the general public as an option which can reduce the national consumption of fossil fuel.

(B) To reduce the fossil fuel consumption and costs of the Federal Government.

(C) To attain the goal of installing solar energy systems in 20,000 Federal buildings by 2010, as contained in the Federal Government's Million Solar Roof Initiative of 1997.

(D) To stimulate the general use within the Federal Government of life-cycle costing and innovative procurement methods.

(E) To develop program performance data to support policy decisions on future incentive programs with respect to energy.

(3) ACQUISITION OF PHOTOVOLTAIC SOLAR ELEC-
TRIC SYSTEMS.—

   (A) IN GENERAL.—The program shall provide for
   the acquisition of photovoltaic solar electric systems and associated storage capability for use in
   public buildings.

   (B) ACQUISITION LEVELS.—The acquisition of
   photovoltaic electric systems shall be at a level substantial enough to allow use of low-cost production techniques with at least 150 megawatts (peak)
   cumulative acquired during the 5 years of the
   program. . . .

Sec. 812. Solar and Wind Technologies.

(a) SOLAR ENERGY TECHNOLOGIES.—The Secretary
shall—

   (1) prepare a detailed roadmap for carrying out the provisions in this title related to solar energy technologies
   and for implementing the recommendations related to
   solar energy technologies that are included in the report transmitted under subsection (e);

   (2) provide for the establishment of 5 projects in geographic areas that are regionally and climatically di-
   verse to demonstrate the production of hydrogen at
   solar energy facilities, including one demonstration
   project at a National Laboratory or institution of
   higher education;

   (3) establish a program—

   (A) to develop optimized concentrating solar power devices that may be used for the production of both electricity and hydrogen; and

   (B) to evaluate the use of thermochemical cycles for hydrogen production at the temperatures attainable with
   concentrating solar power devices;

(4) coordinate with activities sponsored by the Department's Office of Nuclear Energy, Science, and Technology on high-temperature materials, thermochemical cycles, and economic issues related to solar energy;

(5) provide for the construction and operation of new concentrating solar power devices or solar power co-generation facilities that produce hydrogen either concurrently with, or independently of, the production of electricity;

(6) support existing facilities and programs of study related to concentrating solar power devices; and

(7) establish a program—

(A) to develop methods that use electricity from photovoltaic devices for the onsite production of hydrogen, such that no intermediate transmission or distribution infrastructure is required or used and future demand growth may be accommodated;

(B) to evaluate the economics of small-scale electrolysis for hydrogen production; and

(C) to study the potential of modular photovoltaic devices for the development of a hydrogen infrastructure, the security implications of a hydrogen infrastructure, and the benefits potentially derived from a hydrogen infrastructure. . . .

Sec. 25d. Residential Energy Efficient Property.

(a) ALLOWANCE OF CREDIT.—In the case of an individual, there shall be allowed as a credit against the tax imposed by this chapter for the taxable year an amount equal to the sum of—

(1) 30 percent of the qualified photovoltaic property expenditures made by the taxpayer during such year,

(2) 30 percent of the qualified solar water heating property expenditures made by the taxpayer during such year, and

(3) 30 percent of the qualified fuel cell property expenditures made by the taxpayer during such year. . . .

Sec. 179d. Energy Efficient Commercial Buildings Deduction.

(a) IN GENERAL.—There shall be allowed as a deduction an amount equal to the cost of energy efficient commercial building property placed in service during the taxable year.

(b) MAXIMUM AMOUNT OF DEDUCTION.—The deduction under subsection (a) with respect to any building for any taxable year shall not exceed the excess (if any) of—

(1) the product of—

(A) $1.80, and

(B) the square footage of the building, over

(2) the aggregate amount of the deductions under subsection (a) with respect to the building for all prior taxable years. . . .

[Somewhat ironically, this act also failed to extend the production tax credit for solar energy that had been enacted a year earlier in the American Jobs Creation Act of 2004 (as excerpted earlier). The relevant provision was the final section, which lists the renewable energy resources for which production tax credits were eligible.]

Sec. 202. Renewable Energy Production Incentive.

(a) INCENTIVE PAYMENTS.—Section 1212(a) of the Energy Policy Act of 1992 (42 U.S.C. 13317(a)) is amended—

. . .

(b) QUALIFIED RENEWABLE ENERGY FACILITY.—
Section 1212(b) of the Energy Policy Act of 1992 (42
U.S.C. 13317(b)) is amended—

(1) by striking "a State or any political" and all that follows
through "nonprofit electrical cooperative" and inserting "a not-for-profit electric cooperative, a public utility described in section 115 of the Internal Revenue
Code of 1986, a State, Commonwealth, territory, or
possession of the United States, or the District of Columbia, or a political subdivision thereof, an Indian
tribal government or subdivision thereof, or a Native
Corporation (as defined in section 3 of the Alaska Native Claims Settlement Act (43 U.S.C. 1602)),"; and

(2) by inserting "landfill gas, livestock methane, ocean
(including tidal, wave, current, and thermal)," after
"wind, biomass,".

Source: Public Law 109–58.

Solar Energy Farm Use and Potential in the U.S. (2011)

*Agriculture may not be one of the first fields in which applications
of solar energy one thinks of. Yet solar technology has a number of
potential uses in farming, dairying, and other fields of agriculture.
In 2011, the U.S. Department of Agriculture published a detailed
booklet describing many of these uses, with relevant statistics and
examples of solar projects already in use in U.S. agricultural activities. The booklet summarizes the role of solar energy in agriculture
as follows (table references refer to tables included in the report;
footnotes have been omitted):*

Solar energy can supply and/or supplement many farm energy requirements (Table 6). Motor energy generation is the
primary use for PV on farms. Water pumping, one of the

simplest and most prevalent uses of PV, includes irrigation in fields, watering livestock, pond management, and aquaculture. Portable or ground-mounted PV systems can be used to pump water from underground wells or from the surface (e.g. ponds, streams). PV water pumping systems can be the most cost-effective water pumping option in locations where there are no existing power lines. When properly sized and installed, PV water pumps are very reliable and require little maintenance. Environmental benefits can include keeping cattle and other livestock out of wetlands and waterways. The size and cost of a PV water pumping system depends on the local solar resource, pumping depth, water demand, as well as the system purchase and installation costs. Although today's prices for PV panels make most crop irrigation systems expensive, PV systems are very cost effective for remote livestock water supply, small irrigation systems, and pond aeration. While the upfront costs are generally greater than a gas-fuelled [*sic*], generator-based water pumping system, extra costs are met over 5–10 years or sooner in maintenance and fuel cost savings (IREF).

There are a number of other solar applications to be found around the ranch or farm, with the most notable being lighting, electric fencing, battery charging, as well as feeder, sprayer and sprinkler control. PV is an attractive alternative because most applications are considered to be remote and maintenance is easy. Table 7 shows the pricing for a number of on-farm stand-alone applications. Powering buildings is an important application for solar energy on the farm. When grid connection and net metering are available, solar energy can help reduce grid energy needs and balance year-round electricity bills. When a building is off the grid, PV electricity generation provides a good source of energy that can cover needs, especially since running electrical wiring from the grid to an outbuilding can be expensive.

Lighting is another application. Solar can be used for remote building lighting, residential lighting, and large-scale lighting for barns such as hog confinement buildings. Outdoor and security lighting as well as greenhouse lighting are typical off-grid applications. General indoor lighting for farm shops and sheds and lighting for animal production buildings (dairy swine and poultry) may be on or off grid.

Around the farm, solar heat can be used for crop drying instead of the more traditional heating methods with LP gas, electricity, diesel or natural gas. Farmers use a significant amount of energy to dry crops, such as grain, tobacco, and peanuts. Solar heat applications can also be used for livestock and dairy operations. Hog, poultry, and greenhouse farm types have large cooling and space heating loads. Modern hog and poultry farms raise animals in enclosed buildings where it is necessary to carefully control temperature and air quality to maximize the health and growth of the animals. These facilities need to replace the indoor air regularly to remove moisture, toxic gases, odors, and dust. Heating incoming air, when necessary, requires large amounts of energy. With proper planning and design, solar air/space heaters can be incorporated into farm buildings to preheat incoming fresh air. These systems can also induce or increase natural ventilation levels during summer months. Canada's ecoENERGY for Renewable Heat Program, for example, has funded almost 360 poultry barn solar air heating systems. Livestock and dairy operations also have substantial water heating requirements. Solar hot water heating systems can provide hot water for pen cleaning and may be used to supply all or part of hot water requirements in dairy farms. Commercial dairy farms use large amounts of energy to heat water for cleaning milking equipment, as well as to warm and stimulate cow udders. Heating water and cooling milk can account for up to 40% of the energy used on a dairy farm. Aquaculture and breweries are two other industries that can use solar energy for hot water needs.

Source: Xiarchose, Irene M., and Brian Vick. 2011. "Solar Energy Use in U.S. Agriculture: Overview and Policy Issues." U.S. Department of Agriculture. Office of the Chief Economist.

Office of Energy Policy and New Uses. Available online at http://www.usda.gov/oce/reports/energy/Web_SolarEnergy_combined.pdf.

SZ Enterprises, LLC d/b/a Eagle Point Solar v. Iowa Utilities Board, et al. (2014)

One of the legal issues with which advocates of solar energy have to deal is when, how, and under what circumstances an individual or a business can generate electricity from solar energy and then provide that electricity to other consumers in an area. The major reason to impose some restrictions on such activities is that, in nearly all locations, public utility companies have been granted monopoly rights to carry out such businesses. If a company such as Jones Power & Light has been granted such rights, how can, then, some individual or private company be allowed to sell its own solar-produced electricity in the region already served by Jones? An important case dealing with this issue was resolved in 2014, when the Iowa Supreme Court ruled in favor of a small provider against existing power utilities. The court's own description of the case is as follows (some citations omitted at ***):

In this case, we consider whether SZ Enterprises, LLC, d/b/a Eagle Point Solar (Eagle Point) may enter into a long term financing agreement related to the construction of a solar energy system on the property of the city of Dubuque under which the city would purchase from Eagle Point, on a per kilowatt hour (kWh) basis, all of the electricity generated by the system. Prior to proceeding with the project, Eagle Point sought a declaratory ruling from the Iowa Utilities Board (the IUB) that under the proposed agreement (1) Eagle Point would not be a "public utility" under Iowa Code section 476.1 (2011), and (2) Eagle Point would not be an "electric utility" under Iowa Code section 476.22. If Eagle Point was a public utility or an electric utility under these Code provisions, it would be prohibited from serving customers, such as the city, who were located within the exclusive service territory of another electric utility, Interstate Power and Light Company (Interstate Power). ***

The IUB concluded that under the proposed business arrangement, Eagle Point would be a public utility and thus was prohibited from selling the electricity to the city under the proposed arrangement. Because of its ruling on the public utilities question, the IUB found it unnecessary to address the question of whether a party who was not a public utility could nevertheless be an electric utility under the statute.

Eagle Point brought a petition for judicial review. *** The district court reversed. According to the district court, Eagle Point's provision of electric power through a "behind the meter" solar facility was not the type of activity which required a conclusion that Eagle Point was a public utility. The district court further found that although it was conceivable under some circumstances that an entity that was not a public utility could nevertheless be an electric utility under the applicable statutory provisions, Eagle Point's proposed arrangement with the city did not make it an electric utility for purposes of the statutes. The IUB and intervenors MidAmerican Energy Company, Interstate Power, and Iowa Association of Electric Cooperatives, appealed. Eagle Point filed a cross-appeal challenging the reasoning, but not the result, of the district court's electric utility holding.

For the reasons expressed below, we affirm the decision of the district court.

[The court applied an eight-part standard known as the Serv-Yu test in reaching its decision. A key element in its reasoning was as follows:]

The sixth and seventh Serv-Yu factors relate to the ability to accept all requests for service and, conversely, the ability to discriminate among members of the public. *** These twin factors cut in favor of finding that Eagle Point is not a public utility. Eagle Point is not producing a fungible commodity that everyone needs. It is not producing a substance like water that everyone old or young will drink, or natural gas necessary to run the farms throughout the county. More specifically, Eagle Point is not providing electricity to a grid that all may plug into to power their devices and associated "aps," or, more prosaically, their ovens, refrigerators, and lights.

Instead, Eagle Point is providing a customized service to individual customers. Whether Eagle Point can even provide the service will depend on a number of factors, including the size and structure of the rooftop, the presence of shade or obstructions, and the electrical use profile of the potential customer. Further, if Eagle Point decides not to engage in a transaction with a customer, the customer is not left high and dry, but may seek another vendor while continuing to be served by a regulated electric utility. These are not characteristics ordinarily associated with activity "clothed with a public interest."

The eighth Serv-Yu factor is perhaps the most interesting. Under the eighth factor, the actual or potential competition with other corporations whose business is clothed with the public interest is considered. *** Here, the IUB strenuously argues that allowing third-party PPAs will have decidedly negative impacts on regulated electric utilities charged with providing reliable electricity at a fair price to the public. *** The fighting issue in this case is whether factor eight in the Serv-Yu litany trumps the preceding factors and requires that Eagle Point be treated as a public utility providing services to the public.

The position of the IUB has considerable appeal. Certainly, the case can be made that if Eagle Point is allowed to "cream skim" the most profitable customers, there may be impacts on the regulated utility. *** If the third-party-PPA movement gets legs in Iowa, it is conceivable that demand for electricity from traditional utilities will be materially impacted in the long run. . . .

There are also mitigating factors. As pointed out by Eagle Point, it does not seek to replace the traditional electric supplier but only to reduce demand. Although an Eagle Point sale brochure promoting its services is in the record, there is nothing to suggest that its services will be attractive to, or even practical to, many customers of the traditional electric supplier. Further, the parties to third-party PPAs *[power purchase agreements]* have the ability to convert their business arrangements into conventional leases which are outside the scope of regulation. Indeed, in this case, Eagle Point and the city have done just that to avoid unnecessary legal entanglements.

In addition to mitigating factors, there are also countervailing positive impacts. Behind-the-meter solar facilities tend to generate electricity during peak hours when the grid is under the greatest pressure. Further, Iowa Code section 476.8 requires regulated electric utilities to provide reasonably adequate service, and such service must "include programs for customers to encourage the use of energy efficiency and renewable energy sources." Thus, third-party PPAs like the one proposed by Eagle Point actually further one of the goals of regulated electric companies, namely, the use of energy efficient and renewable energy sources. *** In our view, in this case, the balance of factors point away from a finding that the third-party PPA for a behind-the-meter solar generation facility is sufficiently "clothed with the public interest" to trigger regulation. . . .

VI. Conclusion.

For all the above reasons, the decision of the district court is affirmed.

AFFIRMED

Source: In the Supreme Court of Iowa. No. 13–0642. http://www.iowacourts.gov/About_the_Courts/Supreme_Court/Supreme_Court_Opinions/Recent_Opinions/20140711/13-0642.pdf. Accessed on February 4, 2015.

Occupations in Solar Power (2015)

The growth of solar energy in the United States and other parts of the world has opened up a whole new career field that includes many jobs that did not exist a decade ago. The primary source of information about occupations in the United States is the U.S. Bureau of Labor Statistics, which regularly publishes information about the types of jobs available in the United States, the specific job duties involved for each job type, the credentials required, the wages paid, and the prospects in the field. For a complete discussion of careers available in solar energy in the United States, see the reference listed at the end of this section. A general overview of the main types of jobs in the field is provided here from that reference.

Occupations in Scientific Research

Solar power is still gaining popularity and acceptance, so research and development are key aspects of the industry. Continued research and increased returns to scale as production has increased have led to many developments that have decreased costs while increasing efficiency, reliability, and aesthetics. For example, new materials have been developed that allow for low-cost and lightweight thin-film solar panels that are less expensive to produce and easier to transport than glass- or laminate-coated solar panels.

Occupations in scientific research and development have become increasingly interdisciplinary, and as a result, it is common for physicists, chemists, materials scientists, and engineers to work together as part of a team. Most scientists in the solar industry work in an office or laboratory and also spend some time in manufacturing facilities with engineers and processing specialists. . . .

Occupations in Solar Power Engineering

Engineers apply the principles of science and mathematics to develop economical solutions to technical problems. Their work is the link between scientific research and commercial applications. Many engineers specify precise functional requirements, and then design, test, and integrate components to produce designs for new products. After the design phase, engineers are responsible for evaluating a design's effectiveness, cost, reliability, and safety. Engineers use computers extensively to produce and analyze designs, and for simulating and testing solar energy systems. Computers are also necessary for monitoring quality control processes. Computer software developers design the software and other systems needed to manufacture solar components, manage the production of solar panels, and control some solar generating systems.

Most engineers work in offices, laboratories, or industrial plants. Engineers are typically employed by manufacturers of solar equipment and may travel frequently to different worksites, including to plants in Asia and Europe.

Engineers are one of the most sought-after occupations by employers in the solar power industry. According to the Solar Foundation, 53 percent of manufacturing firms reported difficulty in hiring qualified engineers in 2010. . . .

Occupations in Manufacturing for Solar Power

Manufacturing in the solar industry focuses on three technologies: concentrating solar power (CSP), photovoltaic solar power, and solar water heating. However, the vast majority of solar manufacturing firms focus mainly on photovoltaic solar power and producing photovoltaic panels. The production process for photovoltaic panels is more complex than for CSP components, and it involves complicated electronics. Making photovoltaic panels requires the work of many skilled workers, including semiconductor processors, computer-controlled machine tool operators, glaziers, and coating and painting workers. The manufacture of CSP mirrors includes many of the same occupations. . . .

Occupations in Solar Power Plant Development

Building a solar power plant is complex and site selection requires years of research and planning. The proposed site must meet several criteria: large, relatively flat site, adequate sunlight, and minimal environmental impact once built. Prior to beginning construction on a new solar plant, real estate brokers and scientists must ensure the site is suitable and that the proper federal, state, and local permits are obtained for construction of a power plant. . . .

Occupations in Solar Power Plant Construction

Once a site has been selected, civil engineers are responsible for the design of the power plant and related structures. When construction begins, workers are needed to build the actual plant. For a concentrating solar power (CSP) plant, large mirrors are arranged to catch and focus sunlight for power generation, therefore storage tanks, pipes, and generators must be installed before the plant is connected to the electrical grid. Photovoltaic plants are less complex, requiring installation of arrays of photovoltaic panels before they are connected to transformers and the grid. Construction managers have the responsibility of managing the entire construction process. . . .

Occupations in Solar Power Plant Operations

Workers at solar power plants install, operate, and maintain equipment. They also monitor the production process and correct any problems that arise during normal operation. Concentrating solar power (CSP) plants require more workers than photovoltaic plants; photovoltaic plants can sometimes even be run remotely. . . .

Solar Photovoltaic Installers

Solar photovoltaic installers are key to the process of solar panel installation and maintenance. They use specialized skills to install residential and commercial solar projects. They are responsible for safely attaching the panels to the roofs of houses or other buildings and ensuring that the systems work. Solar photovoltaic installers must be able to work with power tools and hand tools at great heights, and possess in-depth knowledge of electrical wiring as well as basic math skills. When necessary, installers must be problem solvers, able to repair damaged systems or replace malfunctioning components. Safety is a priority when installing solar panels because installers run the risk of falling from a roof or being electrocuted by high voltage.

Solar photovoltaic installers are often self-employed as general contractors or employed by solar panel manufacturers or installation companies. Installation companies typically specialize in installing certain types of panels and provide some maintenance and repair services. When a solar panel system is purchased, manufacturers may provide the buyer with installation services or maintenance and repair work. Self-employed installers typically have training and experience with installing solar power systems and are hired directly by the property owners or by a construction firm. . . .

Other Occupations in Solar Panel Installation and Maintenance

Other occupations in solar installation and maintenance are site assessors, electricians, plumbers, and roofers. These workers are involved in the installation process but are not classified as solar photovoltaic installers. However, solar photovoltaic installers possess many of the same skills as these occupations and often have work experience in these fields. . . .

Occupations Supporting the Solar Power Industry

The advancement of the solar power industry has led to job creation in a number of other occupations as well. Many of these jobs do not concentrate on solar power, but they provide support to solar energy production and contribute to the industry as a whole. For instance, the solar power supply chain consists of many different manufacturers of varying sizes. Foundry workers are an important part of this supply chain; they cast metal, plastics, and composites out of raw materials into individual components for solar energy production.

Solar manufacturers need trained salespeople to sell their products to customers. Sales representatives, sales engineers, and sales managers are instrumental in matching a company's products to consumers' needs. They are responsible for making their products known and generating interest in the products. Sales professionals may work directly for manufacturers, distributers, installers, or consulting services. A salesperson must stay abreast of new products and the changing needs of customers. They attend trade shows at which new products and technologies are showcased.

Source: Hamilton, James. 2015. "Careers in Solar Power." Online at http://www.bls.gov/green/solar_power/.

Introduction

Researchers, economists, politicians, and other scholars have been writing about solar energy for well over a century. Today, many thousands of books, articles, reports, and Internet web pages are available on the science and technology of solar power, as well as a host of related issues, including the economics, politics, environmental effects, and other issues associated with the development and use of solar energy. This chapter can list only a small fraction of the many important publications now available on this topic. The chapter lists publications in one of four categories: books, articles, reports, and Internet sites. Some publications are available in more than one form. For example, many scholarly articles on solar power are available both in print and online. In such cases, all sources of a publication are noted in the annotation for the publication.

The majority of items listed in this bibliography are devoted exclusively or primarily to solar energy. However, another whole area of literature exists that discusses renewable and alternative forms of energy as a general category and that almost always includes solar energy as one of those options. That area of books, articles, reports, and Internet sources is very large and is generally not included in this list. For a good source of such resources, see the World Catalog website at www.worldcat.org

A man uses a solar cooker in Zanskar, Ladakh, India. (Baciu/Shutterstock. com)

or the Library of Congress online catalog at http://catalog.loc.
gov/.

Books

Adaramola, Muyiwa, ed. 2014. *Solar Energy: Application, Economics, and Public Perception*. Toronto: Apple Academic Press.
> This book provides a broad general introduction to the
> topic of solar energy with chapters not only on the sci-
> ence and technology of solar power but also additional
> chapters on related topics, such as the economics of solar
> energy and public perceptions of solar power.

Archer, Mary D., and Martin A. Green. 2015. *Clean Electricity
from Photovoltaics*, 2nd ed. London; Hackensack, NJ: Imperial
College Press.
> This collection of articles deals with a variety of technical
> issues involved with the use of solar cells for the production
> of electricity, such as crystalline silicon solar cells, thin-film
> solar cells, polycrystalline cadmium telluride photovoltaic
> devices, and copper-indium-gallium-selenide solar cells.

Barnham, Keith. 2014. *The Burning Answer: A User's Guide to the
Solar Revolution*. London: Weidenfeld and Nicolson.
> This book simply and clearly provides an overview of the
> development of solar power as an energy source, with
> the first chapter devoted to a history of that process, the
> second chapter, to a review of the current status of solar
> power throughout the world, and the third chapter, to a
> prediction as to the possible and probably future of solar
> power as an important element in the world's energy
> equation.

Boxwell, Michael. 2014. *Solar Electricity Handbook: A Simple
Practical Guide to Solar Energy: Designing and Installing Photo-
voltaic Solar Electric Systems*. Ryton on Dunsmore, Warwickshire,
UK: Greenstream Publishing.

This book is designed primarily for individuals who want to build solar power systems that they can use in their own homes and/or businesses. It provides detailed, step-by-step instructions for the construction of virtually every conceivable type of solar power system that can be made for simple nonindustrial, noncommercial settings. The book is also linked to an Internet page that provides answers to and information about additional topics not covered or only partially discussed in the book.

Brown, Lester R. 2015. *Great Transition: Shifting from Fossil Fuels to Wind and Solar Energy.* New York: W. W. Norton.

One of the most influential and prolific writers about alternative energy in the United States, Brown reviews the status of fossil fuels as a source of energy for most of the nation's needs and reviews the benefits that alternative resources, wind and solar in particular, have to offer in the future. He notes the problems that are associated with the use of such resources, and provides responses to those potential issues.

Brownson, Jeffrey R. S. 2014. *Solar Energy Conversion Systems.* Oxford, UK: Academic Press.

The author notes in the introduction to this book that his purpose is "to open up the language of solar energy conversion to a broader audience, to permit discussion of strategies for assessing the solar resource . . . and for designing solar energy conversion systems." Various chapters in the book focus on topics such as the laws of light; physics of light, heat, work, and photoconversion; meteorology: the many facets of the sky; Sun–Earth geometry; applying the angles to shadows and tracking; solar energy economics; and solar project financing.

Butti, Ken, and John Perlin. 1980. *A Golden Thread: 2500 Years of Solar Architecture and Technology.* Palo Alto: Cheshire Books; New York: Van Nostrand Reinhold.

This book is one of the classic studies of the history of solar energy and the way it has been put to use by humans for two-and-a-half millennia. For a companion volume with more recent information, see Perlin.

Chwieduk, Dorota. 2014. *Solar Energy in Buildings: Thermal Balance for Efficient Heating and Cooling.* San Diego, CA: Academic Press.

This book is intended to provide an analysis and description of the way in which solar energy can be facilitated in the construction of new buildings. Chapters focus on topics such as the fundamentals of solar radiation, the availability of solar energy on Earth's surface, shaping the building envelope to take maximum advantage of solar energy available in an area, passive and active solar energy systems, photothermal conversion within a given building, and energy balances within a structure.

Eicker, Ursula. 2014. *Energy Efficient Buildings with Solar and Geothermal Resources.* Chichester, West Sussex, UK: John Wiley & Sons.

This book is intended for engineers and students who would like to know more about the use of solar and geothermal power in meeting the energy needs of a building. It discusses topics such as energy consumption of buildings, photovoltaics, solar heating and cooling, and geothermal heating and cooling.

Flournoy, Don M. 2012. *Solar Power Satellites.* New York: Springer.

This book provides an excellent and comprehensive review of all aspects of the development and use of satellites to collect and transmit solar energy to Earth for commercial use on the planet. Individual chapters deal with topics such as the technical description of solar satellites, issues involved with the maintenance and use of such devices, launch problems and issues, and challenges faced by use of solar satellites.

Fraas, Lewis M. 2014. *Low-cost Solar Electric Power*. Cham, Switzerland: Springer.

This book provides a very comprehensive overview of the use of solar energy for electrical production, beginning with a review of the development of solar power and the scientific principles on which solar-to-electrical energy production is based. It then discusses the range of technologies available for the production of electricity from solar energy, including traditional photovoltaics, thin-film photovoltaics, multi-junction solar cells, and infrared and thermal photovoltaics. The final chapter deals with the production of electricity from space mirrors.

Gonzalez, George A. 2012. *Energy and Empire: The Politics of Nuclear and Solar Power in the United States*. Albany: State University of New York Press.

Nuclear and solar power are two of the alternatives frequently mentioned in discussions of modalities available for dealing with a new energy world with diminished supplies of coal, oil, and natural gas. This book reviews the potential of these two forms of energy and examines the political, social, economic, and other forces that have been and will be brought to bear in making decisions about the employment of these two technologies in the future.

Hamilton, James. 2011. *Careers in Solar Power*. Washington, DC: U.S. Bureau of Labor Statistics. http://www.bls.gov/green/solar_power/. Accessed on February 2, 2015.

This booklet is part of the Bureau of Labor Statistics' series on careers in various occupations. It provides background information such as job duties, credentials, educational requirements, wages, and projected opportunities for various types of jobs in the solar power industry.

Irvine, Stuart J. C., ed. 2015. *Materials Challenges: Inorganic Photovoltaic Solar Energy*. Cambridge: Royal Society of Chemistry.

The essays that make up this volume review the current status of the types of materials being used in photovoltaic cells, such as thin-film silicon, cadmium telluride, chalcogenides, and nanomaterials, and possible future applications of these and related materials.

Jordan, Philip G. 2014. *Solar Energy Markets: An Analysis of the Global Solar Industry*. London: Elsevier.

This book provides an excellent overview of the current status of solar power worldwide. Following an introductory chapter on the mechanics of solar power, the book turns to a consideration of "the new culture of environmentalism" in the United States and other parts of the world that contributes to the growth of interest in solar energy, finance and venture capital opportunities in the field of solar energy, global policies on solar energy, the solar labor market, and the economics of solar energy.

Kennedy, Danny. 2012. *Rooftop Revolution: How Solar Power Can Save Our Economy and Our Planet from Dirty Energy*. San Francisco: Berrett-Koehler Publishers.

This book provides a useful introduction to the subject of solar energy and its potential for helping humans deal with energy issues in the future. In addition to providing a technical introduction to the topic, the author also provides a review of some of the political issues involved in financing and supporting research on solar energy, some important economic issues, and the ways solar energy can be a factor in a person's personal life.

Kulichenko, Natalia, and Jens Wirth. 2012. "Concentrating Solar Power in Developing Countries: Regulatory and Financial Incentives for Scaling Up." Washington, DC: World Bank.

Most developing nations are severely handicapped in their plans for development by inadequate energy resources of any kind, whether they be traditional fossil fuels that have powered the growth of developed nations, or alternative energy sources, which have seldom been developed by developing economies. This publication from the World Bank examines the potential of solar power for providing the energy resources needed by developing nations, the factors that have hindered such development in the past, and the decisions that developing nations can take to improve their use of solar energy as a part of their future energy equations.

Li, Hejun. 2014. *China's New Energy Revolution How the World Super Power Is Fostering Economic Development and Sustainable Growth through Thin-Film Solar Technology.* New York: McGraw-Hill.

The Chinese government has made a firm commitment to the development of solar power as a significant part of the nation's energy equation in the future. This book describes the background to that decision, the type of solar power systems that are being developed, and how solar power is expected to be an essential part of the Chinese economy in the future.

Lindsay, Porter. 2015. *The Renewable Energy Home Handbook Insulation & Energy Saving, Living Off-Grid, Bio-Mass Heating, Wind Turbines, Solar Electric PV Generation, Solar Water Heating, Heat Pumps, & More.* Poundbury, Dorset, UK: Veloce Publishing.

As the title of this book suggests, the author provides information on virtually every conceivable aspect of the ways in which solar power can be used in residential settings, including all of the situations and technologies mentioned in the title.

Maeda, Martha. 2011. *How to Solar Power Your Home: Everything You Need to Know Explained Simply.* Ocala, FL: Atlantic Publishing Group.

As the book title suggests, Maeda provides the information a homeowner needs to decide whether or not to invest in solar energy systems for their own homes and, if so, what steps are involved in acting on that decision.

Mankins, John. 2014. *The Case for Space Solar Power*. Houston: Virginia Edition Publishing.

The author begins by noting that most discussions of solar energy today focus almost entirely on Earth-based systems, but that space-based devices have a number of important advantages over such systems. He describes the technology, economics, politics, and other aspects of space-based solar systems and argues that they are able to provide a significant contribution to solving the world's energy needs in the future.

Mattson, Brad. 2014. *The Solar Phoenix: How America Can Rise from the Ashes of Solyndra to World Leadership in Solar 2.0*. Los Gatos, CA: Robertson Publishing.

The author explains that the United States has completed stage Solar 1.0 by passing the point where solar energy is now economically and technically competitive with traditional energy sources, such as coal, oil, and gas, and that the nation should now use the technical know-how that it has developed to lead the rest of the world in beginning to produce and develop solar energy as the solution for many of its most fundamental problems.

McKevitt, Steve, and Anthony J. Ryan. 2014. *The Solar Revolution: One World, One Solution, Providing the Energy and Food for 10 Billion People*, rev ed. London: Icon Books.

This book makes grandiose claims for the role that solar energy can play in solving many of the world's most urgent problems, such as providing food, transportation, and industrial operations for a planet with an ever-increasing population. Various chapters of the book explain how solar energy can be implemented in problem solving in each of these areas.

Mislin, H., and R. Bachofen, eds. 2014. *New Trends in Research and Utilization of Solar Energy through Biological Systems*. Basel, Switzerland: Birkhäuser.

> The 30 chapters in this book review recent developments in a variety of methods for converting solar energy by using biological systems. Some sample topics are "Sugar crops as a solar energy convertor," "Plants as a direct source of fuel," "Leaf protein as a food source," "The ocean as a supplier of food and energy," "Mass production of micro-algae," and "Photosynthetic production of ammonia."

Nakaya, Andrea C. 2013. *What Is the Future of Solar Power?* San Diego, CA: ReferencePoint Press.

> This book is intended for young adults, grade eight and up. It is part of the publisher's *The Future of Renewable Energy* series that discuss the pros and cons of various types of renewable energy for meeting the world's future energy needs.

Norton, Brian. 2014. *Harnessing Solar Heat*. Dordrecht, the Netherlands: Springer.

> This book is intended as a textbook that presents and explain fundamental concepts in the use of solar energy for the production of heat. It includes topics such as the properties of solar radiation in general, optics and heat transfer in solar collectors, storage of solar heat, greenhouses, and active and passive solar systems in buildings.

Palz, Wolfgang, ed. *Solar Power for the World: What You Wanted to Know about Photovoltaics*. Singapore: Pan Stanford Publishing.

> This collection of papers is biased heavily toward discussions of the historical development of solar power in the United States and other parts of the world. Some topics include "Looking Back to Light the Future," "Solar Power for Space Satellites," "First Ideas about Lighting with Solar Power," "The Pioneering Role of the United

States," "The Ethical Implications of Photovoltaics," and "The Role of Stakeholders in Society."

Pedersen, Peder Vejsig, et al. 2015. *Green Solar Cities*. New York: Routledge.

> The authors describe a project conducted in Copenhagen that attempts to vastly increase the use of solar energy for the energy needs of buildings constructed in the city. Their goal in describing this project is to attempt to influence construction companies, building materials suppliers, governmental officials in other nations, and other individuals and organizations responsible for the design and construction of urban communities.

Perlin, John. 2013. *Let It Shine: The 6,000-Year Story of Solar Energy*. Novato, CA: New World Library.

> This excellent historical account of the history of solar energy is divided into six main chapters: early use of the Sun (from about 6000 BCE to about 1800), power from the Sun (from 1860 to 1914), solar water heating (from 1891 to the Second World War), solar house heating (from 1807 to 1962), photovoltaics (1876 to 1968), and the post–oil embargo era (1945 to the present day).

Plante, Russell H. 2014. *Solar Energy: Photovoltaics and Domestic Hot Water, a Technical and Economic Guide for Project Planners, Builders, and Property Owners*. Amsterdam; Boston: Academic Press.

> The title of this book describes well its primary purpose. Individual chapters deal with topics such as basic science and technology of solar power, the use of solar power for the production of hot water in domestic systems, and the economics of solar domestic hot water systems. Three appendices provide useful charts, tables, and calculators for determining the parameters and economic aspects of specific domestic hot water systems.

Razeghifard, Reza, ed. 2013. *Natural and Artificial Photosynthesis: Solar Power as an Energy Source.* Hoboken, NJ: John Wiley and Sons.

Photosynthesis is, of course, probably the oldest single method for capturing solar energy on planet Earth. For some time now, researchers have been trying to understand that process more completely so that they could then develop artificial photosynthetic technologies that could be used to generate energy for a wide variety of synthetic processes. The papers in this book discuss processes such as photosynthetic water splitting, a variety of artificial photosynthetic technologies, use of bacteria and algae in photosynthetic processes, and photobioreactors.

Richter, Christoph, Daniel Lincot, and Christian A. Gueymard, eds. 2013 *Solar Energy.* New York: Springer.

This anthology collects 30 articles from the *Encyclopedia of Sustainability Science and Technology*, dealing with a number of topics in the field of solar energy, such as photovoltaics, solar thermal energy, and solar radiation.

Rule, Troy A. 2014. *Solar, Wind and Land: Conflicts in Renewable Energy Development.* London; New York: Routledge, Taylor & Francis Group.

Interest in the development and implementation of renewable energy resources, such as wind, water, solar, and geothermal, has increased dramatically in the last few decades, and is likely to become even more important in the nation's and world's energy equation in the future. Yet, these promising technologies all have certain disadvantages associated with them, one of the most common of which is land use issues. For example, both wind and solar facilities require very large areas of land for their installation and operation. This book reviews land issues created in particular by the development of wind and solar power systems.

Schaeffer, John. 2015. *Real Goods Solar Living Sourcebook: Your Complete Guide to Living Beyond the Grid with Renewable Energy Technologies and Sustainable Living*, 14th ed. Hopland, CA: Real Goods Solar Living Center; Gabriola Island, BC: New Society Publishers.

> The author of this very popular book directs the work at individuals who wish to begin using or to extend their use of solar power for home and residential purposes. He provides a host of valuable background information about solar energy as well as detailed instructions about its use in residential and commercial settings.

Smith, Mervyn, James Russell, and Tony Milanowski. 2014. *Solar Energy in the Winemaking Industry*. London: Springer.

> This interesting book describes in considerable detail the use of solar power systems in one specific industry, winemaking, with a detailed discussion of the process of winemaking and its ancillary activities (such as economic issues) and the way in which solar energy systems have been integrated into the industry. A number of helpful case studies are provided.

Varadi, Peter F. *Sun above the Horizon: Meteoric Rise of the Solar Industry*. Singapore: Pan Stanford Publishing.

> The author divides the history of solar power into three sections, Dawn, from 1972 to 1984; Sunrise, from 1985 to 1999; and Towards High Noon, from 2000 to 2013. The book provides an excellent explanation for the general reader of the changes in science, technology, politics, economics, and politics that have been involved in the development of solar power over the last decades.

Williams, Neville. 2014. *Sun Power: How Energy from the Sun Is Changing Lives around the World, Empowering America, and Saving the Planet*. New York: Forge Books.

> This book is intended for a general audience. It provides a very nice history of the development of solar power and

can be of interest in the use of solar energy by way of a dozen events that have occurred primarily in the United States since 1979 and beginning with the story of the formation of the Paradise Power Company of Taos, New Mexico, in that year.

Articles

Some of the journals dealing primarily with solar energy are the following: *Applied Solar Energy (*ISSN: 0003-701X*)*; *International Journal of Renewable Energy Research* (ISSN: 1309-0127); *Journal of Solar Energy Engineering* (ISSN: 0199-6231); and *Solar Energy* (ISSN: 0038-092X).

Adey, Walter H., Patrick C. Kangas, and Walter Mulbry. 2011. "Algal Turf Scrubbing: Cleaning Surface Waters with Solar Energy While Producing a Biofuel." *BioScience.* 61(6): 434–441. http://enst.umd.edu/sites/default/files/_docs/BioScience%20Article.pdf. Accessed on March 31, 2015.

> The authors describe a system that uses solar energy to force wastewater over a surface covered with algae that results in the production of clean water and from which the algae can then be harvested for the production of biofuels.

Aman, M. M., et al. 2015. "A Review of Safety, Health and Environmental (SHE) Issues of Solar Energy System." *Renewable and Sustainable Energy Reviews.* 41: 1190-1204.

> Solar energy is frequently praised for its many environmental benefits and its relative safety for human health. However, some health issues *are* associated with the construction and operation of solar facilities. This paper reviews current understanding of those health issues.

Baharoon, Dhyia Aidroos, et al. 2015. "Historical Development of Concentrating Solar Power Technologies to Generate Clean Electricity Efficiently—a Review." *Renewable and Sustainable Energy Reviews.* 41: 996-1027.

The authors summarize some of the major technologies that have been developed for the collection and conversion of solar energy to determine which of those technologies appears to be most efficient and best able to integrate with existing fossil fuel technologies during a period of transition from convention to alternative fuels for electrical generation.

Bronin, Sara C. 2009. "Solar Rights." *Boston University Law Review.* 89(4): 1217-1266. https://www.bu.edu/law/central/jd/organizations/journals/bulr/documents/BRONIN.pdf. Accessed on March 31, 2015.

This article provides a comprehensive review of the legal issues related to the question of an individual's rights to access to solar radiation for use in his or her solar roof panels or other solar facility.

Brown, Marilyn. 2015. "Innovative Energy-Efficiency Policies: An International Review." *Wiley Interdisciplinary Reviews: Energy and Environment.* 4(1): 1–25.

The author provides a comprehensive review of the types of energy-efficient programs and policies that have been developed at various locations around the world, including the use of various types of solar power. She suggests that the main value of this type of review is to make sure that "failures of the past can be avoided and successes can be replicated and expanded."

Burnett, Dougal, Edward Barbour, and Gareth P. Harrison. 2014. "The UK Solar Energy Resource and the Impact of Climate Change." *Renewable Energy.* 71: 333–343.

The authors report on their efforts to calculate the solar energy available to the United Kingdom and how predicted climate change is likely to change that number. They conclude that climate change is likely to increase the amount of solar energy available, although with considerable seasonal variation and discrepancies among various geographical regions.

Curreli, Alessandra, and Helena Coch Roura. 2013. "Urban Layout and Façade Solar Potential: A Case Study in the Mediterranean Region." *ACE: Architecture, City and Environment.* 7(21): 117–132. http://upcommons.upc.edu/revistes/bitstream/ 2099/13010/7/ACE_21_SA_14.pdf. Accessed on March 30, 2015.

This paper provides a fascinating and detailed description of the factors involved in building a community that makes optimal use of passive solar power in its design.

Edwards, Peter P., et al., eds. 2013. "Can Solar Power Deliver?: Papers of a Discussion Meeting." *Philosophical Transactions.* 371(1996): whole.

This issue of the journal *Philosophical Transactions* is devoted to a series of papers presented at a meeting held at the Royal Society in November 2011 to discuss this question. The papers in the collection cover topics such as the future of concentrated solar power, the use and misuse of photosynthesis as a tool for solar energy production, the state-of-the-art silicon solar cells, concentrating solar thermal power, and realizing the potential of solar power in the European Union.

Ernst, Deliana. 2013. "Beam It down, Scotty: The Regulatory Framework for Space-Based Solar Power." *Review of European Community and International Environmental Law.* 22(3): 354-365.

The author points out that suggestions for space-based solar systems have been around for more than half a century and that such systems could begin to operate on an economically efficient basis within the next three decades. She notes that that possibility may raise some serious fundamental legal questions about the operation and control of space- and Earth-based solar systems. She reviews relevant existing laws and regulations dealing with this issue.

Grätzel, Michael. 2001. "Molecular Photovoltaics That Mimic Photosynthesis." *Pure and Applied Chemistry*. 73(Part 3): 459–468.
One of the inventors of the dye-sensitized solar cell provides a broad general introduction to the concept of artificial photosynthesis as a mechanism for producing electrical current from solar radiation.

Hermerschmidt, Felix, et al. 2015. "Beyond Solar Radiation Management—the Strategic Role of Low-cost Photovoltaics in Solar Energy Production." *International Journal of Sustainable Energy*. 34(3-4): 211–220.
This very optimistic paper outlines the potential role of solar power in future energy systems, suggesting that "PVs [photovoltaics] will be the most significant electricity source if the cost per kWh produced is further reduced." The authors use Cyprus as an example of a nation in which the role of solar energy has increased over recent decades and can expand even further if specific actions are taken to support the technology.

Hernandez, R.R., et al. 2014. "Environmental Impacts of Utility-Scale Solar Energy." *Renewable and Sustainable Energy Reviews*. 29: 766–779.
In this review paper, the authors consider the direct and indirect environmental impacts, both beneficial and adverse, of utility-scale solar energy (USSE) development, such as impacts on biodiversity, land-use and land-cover change, soils, water resources, and human health.

Hirth, Lion. 2015. "Market Value of Solar Power: Is Photovoltaics Cost-Competitive?" *IET Renewable Power Generation*. 9(1): 37–45.
One of the key questions relating to solar power is whether it is economically competitive with more traditional forms of power, such as coal- and oil-powered electrical

generation. A number of other questions are related to this basic issue, such as how one determines the economic "cost" of power generation. This article attempts to answer the question about the economic viability of solar power as of early 2015.

Humphreys, Gary. 2014. "Harnessing Africa's Untapped Solar Energy Potential for Health." *Bulletin of the World Health Organization.* 92(2): 82–83. http://www.who.int/bulletin/volumes/92/2/14-020214.pdf. Accessed on January 30, 2015.

This article points out that individuals and organizations in Africa make very little practical use of the very large solar energy resource enjoyed by the continent. It then goes on to describe a simple new device that can be used in local settings to convert solar energy to electricity for a variety of health-related issues.

Jakubowski, Greg. 2014. "Solar Response: Fighting Fires around Solar Power Systems." FireRescue. http://www.firefighternation.com/article/firefighting-operations/tackling-solar-power-challenges. Accessed on March 30, 2015.

This article deals with a topic of specialized interest, fire fighting in areas around solar power systems. But it provides an interesting insight into the issues faced by first responders to the growing challenge of protecting such systems from a variety of disasters.

Kasaeian, Alibakhsh, Amin Toghi Eshghi, and Mohammad Sameti. 2015. "A Review on the Applications of Nanofluids in Solar Energy Systems." *Renewable and Sustainable Energy Reviews.* 43: 584–598.

The authors review the ways in which nanofluids can be used in solar power systems to increase the efficiency of such systems as well as reduce their environmental effects.

Khalil, Hafiz Bilal, and Syed Jawad Hussein Zaidi. 2014. "Energy Crisis and Potential of Solar Energy in Pakistan." *Renewable and Sustainable Energy Reviews.* 31: 194-201.

A number of papers are available in the literature discussing the role of solar power in specific nations and regions around the world. This report is an example of those papers. It points out that Pakistan is an "energy-deficient" nation, examines the reason that such is the case, and suggests ways in which solar energy can be used to meet the nation's future energy needs.

Komiyama, Ryoichi, and Yasumasa Fujii. 2015. "Analysis of Japan's Long-Term Energy Outlook Considering Massive Deployment of Variable Renewable Energy under Nuclear Energy Scenario." *Electrical Engineering in Japan*. 190(2): 24–40.

Lacking any significant amount of fossil fuel resources, Japan has long relied heavily on the development of nuclear power to meet its energy needs. The Fukushima disaster of 2011, however, has forced the Japanese government to rethink its view of the place of nuclear power in the nation's economy and place greater emphasis on alternative sources of energy, primarily wind and solar. This article considers the changes that may be necessary to decommission all nuclear plants by the year 2050.

Lee, Kyoung-ho, et al. 2014. "Annual Measured Performance of Building-Integrated Solar Energy Systems in Demonstration Low-energy Solar House." *Journal of Renewable and Sustainable Energy*. v(4): 042013. (Online article). http://scitation.aip.org/content/aip/journal/jrse/6/4/10.1063/1.4893467. Accessed on January 29, 2015.

The authors report on an energy audit for a house built on the campus of the Korea Institute of Energy Research designed to operate almost entirely on solar energy. They provide data on the efficiency of both integrated solar collectors and photovoltaic modules installed in the house.

Liu, Jia, et al. 2014. "A Solar Energy Storage and Power Generation System Based on Supercritical Carbon Dioxide." *Renewable Energy*. 64: 43–51.

The authors describe a new type of solar energy conversion and storage system that uses supercritical carbon dioxide rather than water as its working fluid, resulting in a system that is substantially more efficient than the traditional system.

Lovich, Jeffrey E., and Joshua R. Ennen. 2011. "Wildlife Conservation and Solar Energy Development in the Desert Southwest, United States." *BioScience*. 61(12): 982–992. https://profile.usgs.gov/myscience/upload_folder/ci2011 Dec3008522333446bio%20science.2011.61.12.8.pdf. Accessed on March 31, 2015.

The authors point out that there are a number of potential hazards posed by solar energy facilities in the U.S. Southwest, including "habitat fragmentation and barriers to gene flow, increased noise, electromagnetic field generation, microclimate alteration, pollution, water consumption, and fire." They conclude that there are insufficient data at this point to determine the severity of these potential hazards to wildlife in the region and, therefore, encourage more studies on the issue.

MacNeill, Maureen. 2014. "Solar Energy: Using the Sun's Heat to Get Out Oil." *OPEC Bulletin*. 45(3): 16–23.

MacNeill explains how solar energy can be used to improve the efficiency of oil exploration and extraction processes, and how some of the earliest models of that process are now being tried out in Kuwait.

Marcelino, R.B.P., et al. 2015. "Solar Energy for Wastewater Treatment: Review of International Technologies and Their Applicability in Brazil." *Environmental Science and Pollution Research*. 22(2): 762–773.

The authors note that conventional water treatment facilities are often inadequate for the neutralization of certain types of industrial wastes, but that another type of treatment using solar power known as advanced oxygen processes (AOP) is able to solve this problem. They suggest that the abundance of solar radiation in Brazil makes the use of AOPs for wastewater treatment an attractive option for such plants in that country.

Marshall, Jessica. 2014. "Solar Energy: Springtime for the Artificial Leaf." *Nature*. 510(7503): 22–24.

This article reviews the progress of research in the field of artificial photosynthesis as a means for capturing solar energy and converting it into electricity for transportation systems. The types of devices used in such systems are often referred to as "artificial leaves."

Mehrdadi, N., et al. 2007. "Aplication *[sic]* of Solar Energy for Drying of Sludge from Pharmaceutical Industrial Waste Water and Probable Reuse." *International Journal of Environmental Research*. 1(1): 42–48.

This article provides a description of one possible, and somewhat unusual, use of solar energy in industry, as a way of solving a common waste problem of the pharmaceutical industry. http://www.bioline.org.br/pdf?er07007. Accessed on March 30, 2015.

Nazeeruddin, Md. K., Etienne Baranoff, and Michael Grätzel. 2011. "Dye-Sensitized Solar Cells: A Brief Overview." *Solar Energy*. 85(6): 1172–1178.

The aim of this paper is "to give a short and simple overview of the dye-sensitized solar cell technology from the working principles to the first commercial applications." The paper is clearly written and readily accessible to most nonprofessionals interested in the field of solar energy.

Neale, Nathan R. 2014. "Solar Energy: Packing Heat." *Nature Chemistry*. 6(5): 385–386.

This review article discusses research on systems that can capture solar radiation and store it in chromophores, after which it can later be released under circumstances in which it is needed.

Pan, Shaowu, et al. 2014. "Wearable Solar Cells by Stacking Textile Electrodes." *Angewandte Chemie International.* 53(24): 6110–6114.

    The authors describe their research on fabrics into which are woven micrometer-sized metal wires and nanotubes that act as electrodes that are activated by solar radiation. The electrical energy produced by the system can be used to power or recharge a variety of small electrical devices, such as one's cell phone.

Parkab, Sang-Chul, et al. 2010. "Alternative Energy Policies in Germany with Particular Reference to Solar Energy." *Journal of Contemporary European Studies.* 18(3): 323–320.

    This article provides an excellent general overview of the policy with regard to solar energy that has been developed in Germany, to some extent as a reflection of the attitudes of other nations in the European Union about solar energy.

Sarralde, Juan José, et al. 2015. "Solar Energy and Urban Morphology: Scenarios for Increasing the Renewable Energy Potential of Neighbourhoods in London." *Renewable Energy.* 73: 10–17.

    This very interesting paper describes the ways in which an expanded use of solar power would change the physical features and other characteristics of neighborhoods in the city of London where such facilities were installed.

Sharon, H., and K. S. Reddy. 2015. "A Review of Solar Energy Driven Desalination Technologies." *Renewable and Sustainable Energy Reviews.* 41: 1080–1118.

    One of the many proved applications of solar energy is in the desalination of water. This paper reviews some of the processes that are available for the use of solar energy for the desalination of salt and saline water.

Trieb, Franz, Tobias Fichter, and Massimo Moser. 2014. "Concentrating Solar Power in a Sustainable Future Electricity Mix." *Sustainability Science*. 9(1): 47–60.

The authors review some of the properties of solar plants now and the future and assess how they can be integrated into energy-generating systems in the future. They say that solar plants are currently economically competitive with oil-fired electricity-generating plants, and predict that they will be similarly competitive with natural gas plants by 2020 and with coal-powered plants by 2025.

Trieb, Franz, et al. 2012. "Solar Electricity Imports from the Middle East and North Africa to Europe." *Energy Policy*. 42: 341–353.

The authors attempt to estimate the feasibility of constructing concentrating solar power facilities in North Africa and the Near East and shipping that power produced to Europe. They find that the project is feasible with "less than 0.2% of the land suitable for CSP plants [in North Africa and the Near East] would be enough to supply 15% of the electricity demand expected in Europe in the year 2050."

Turney, Damon, and Vasilis Fthenakis. 2011. "Environmental Impacts from the Installation and Operation of Large-Scale Solar Power Plants." *Renewable and Sustainable Energy Reviews*. 15(6): 3261–3270.

The authors point out that the increasing number of solar power facilities being built requires very large areas of land for their installation. As a result, a number of environmental concerns have been raised about the construction and operation of such plants. They select 32 such concerns in the areas of land use, human health and well-being, plant and animal life, geohydrological resources, and climate change to study. They find that the environmental impact of solar power plants is beneficial in 22 of those instances, four are neutral, and six require further research before their effect can be assessed.

Williges, Keith, John Lilliestam, and Anthony Patt. 2010. "Making Concentrated Solar Power Competitive with Coal: The Costs of a European Feed-In Tariff." *Energy Policy*. 38(6): 3089–3097.

One of the proposals for dealing with future energy needs in the European Union is to build solar plants in North Africa and transmit the electricity produced in those plants to Europe by extensive transmission systems. This paper examines the kind and amount of government subsidies that would be needed to make such a program economically viable and concludes that such costs are "reasonable for the EU." See also the article by Trieb et al. on this issue.

Yu, Jian. 2014. "Bio-Based Products from Solar Energy and Carbon Dioxide." *Trends in Biotechnology*. 32(1): 5–10.

The author suggest using a process involving solar energy and carbon dioxide as a more direct way of producing energy from sunlight rather than the more complex and roundabout process of biomass conversion.

Reports

Committee on U.S.-China Cooperation on Electricity from Renewable Resources; National Research Council; Chinese Academy of Sciences; Chinese Academy of Engineering. 2010. "The Power of Renewables." Washington, DC: National Academies Press. https://www.nap.edu/login.php?record_id=12987&page= http%3A%2F%2Fwww.nap.edu%2Fdownload.php%3F record_id%3D12987. Accessed on March 31, 2015.

This report focuses on the future of renewable energy production in the world's two largest economies, the United States and China, with regard to energy use, supply, and demand; energy resources; and policy implications.

"Deutsche Bank's 2015 Solar Outlook: Accelerating Investment and Cost Competitiveness." 2015. Deutsche Bank. https://www.db.com/cr/en/concrete-deutsche-banks-2015-solar-outlook.htm. Accessed on March 30, 2015.

> Deutsche Bank's 2015 study of solar electricity suggests that the technology will be economically competitive with other forms of electricity generation by 2016 in 47 of the 50 states (full report available only by subscription).

Farhar-Pilgrim, Barbara. 2010. "Community Response to Concentrating Solar Power in the San Luis Valley: October 9, 2008—March 31, 2010." Golden, CO: National Renewable Energy Laboratory.

> This report summarizes a very interesting study as to how stakeholders in the San Luis Valley of California thought about and responded to the installation and operation of a utility-sized solar power facility in their region. The researchers found that people in the area generally knew relatively little about solar power, but were, nonetheless, generally supportive of its use in contributing to the area's energy equation.

Hand, M.M., et al., eds. 2012. "Renewable Energy Futures Study." 4 vols. Golden, CO: National Renewable Energy Laboratory. http://www.nrel.gov/analysis/re_futures/. Accessed on March 31, 2015.

> In many ways, this report provides the most complete overview of the history, current status, and future potential of solar energy (as well as other forms of renewable energy) currently available. The text contains both extensive technical details and a comprehensive review of solar energy easily comprehensible to the average reader.

International Energy Agency. 2010. "Technology Roadmap: Concentrating Solar Power." Paris: OECD/IEA.

> This report is part of a series of publications by the International Energy Agency on the development of alternative

and renewable forms of energy needed to meet human demands for energy in the future. The report covers topics such as the status of concentrating solar power (CSP) systems today, a vision of future deployment of such systems, economic perspectives on the future of solar power, milestones for technological development, and a policy framework for the expansion of CSP systems.

International Energy Agency. 2014. "Technology Roadmap: Solar Photovoltaic Energy." http://www.iea.org/publications/ freepublications/publication/TechnologyRoadmapSolarPhotovol taicEnergy_2014edition.pdf. Accessed on March 31, 2015.

International Energy Agency. 2014. "Technology Roadmap: Solar Thermal Electricity." http://www.iea.org/pub lications/freepublications/publication/TechnologyRoad mapSolarThermalElectricity_2014edition.pdf. Accessed on March 31, 2015.

These reports are part of a series of publications by the International Energy Agency on the development of alternative and renewable forms of energy needed to meet human demands for energy in the future. The reports cover topics such as progress in the field since the last such report in 2009, visions for deployment of additional solar PV and solar thermal electric facilities, developments in solar PV technology and solar thermal electricity, system integration of solar PV electricity and solar thermal electricity, policy and financial considerations, and roadmap plans for the next four decades for each type of solar technology.

Morley, David. 2014. "Planning for Solar Energy: Promoting Solar Energy Use through Local Policy and Action." Chicago: APA Planners.

This report (PAS 575) provides suggestions for local communities who wish to make use of solar energy for residential, commercial, electricity-generating, or other purposes. The report is available to the general public free of charge at https://www.planning.org/store/product/?ProductCode=BOOK_P575.

Stoffel, Tom, et al. 2010. "Concentrating Solar Power: Best Practices Handbook for the Collection and Use of Solar Resource Data." National Renewable Energy Laboratory. Technical Report NREL/TP-550–47465. http://www.nrel.gov/docs/fy10o sti/47465.pdf. Accessed on February 2, 2015.

This report was prepared by the National Renewable Energy Laboratory in response to a perceived need "for a single document addressing the key aspects of solar resource characterization." The document is an excellent source of basic information about many aspects of the collection of solar energy and its use in the generation of electrical energy and other applications.

UNIDO International Solar Energy Center for Technology Promotion and Transfer (UNIDO-ISEC), and Jiangsu Modern Low-Carbon Technology Research Institute. 2014. *The Research Report on Application of Low-Carbon Technology in Expo 2010 Shanghai*. Heidelberg, Germany: Springer.

This book offers a summary of papers presented, exhibitions shown, and issues discussed at the Expo 2010 Shanghai conference on low-carbon technologies. The first section of the report consists of a general overview of the issues raised at the conference. The second section focuses on a number of low-carbon technologies discussed at the conference, such as solar, wind, and biomass energy. The third section of the book describes some of the specific exhibits presented at the meeting, including those of the Swiss, Italian, Finnish, and Canadian pavilions. The final section offers some examples of "low-carbon life" exhibited at the conference, such as milk carton benches, low-carbon garments, and green bicycles.

Xiarchos, Irene, and Brian Vick. "Solar Energy Use in Agriculture: Overview and Policy Issues." 2011. Office of Energy Policy

and New Uses. U.S. Department of Agriculture. http://www
.usda.gov/oce/reports/energy/Web_SolarEnergy_combined.pdf.
Accessed on March 30, 2015.

    This report discusses past and current uses of solar energy
in agricultural operations, along with future potential for
the technology. Various chapters deal with the basics of
solar energy as well as selected case studies, financial considerations, and federal policies on solar energy that are
relevant to agricultural applications.

Internet Sources

A number of online websites are available carrying up-to-date
news and information about developments in solar energy. The
following are a sample of those websites:

Alternative Energy *News*: http://www.alternative-energy-news.
info/headlines/solar/

Electric Light and Power: http://www.elp.com/topics/solar-en
ergy.htm

Greentech Media: http://www.greentechmedia.com/articles/cate
gory/solar

Nature: http://www.nature.com/nature/supplements/collections/
energy/

The New York Times: http://topics.nytimes.com/top/news/busi
ness/energy-environment/solar-energy/index.html

Renewable Energy. World.com: http://www.renewableenergy
world.com/rea/home/solar-energy

Science Daily: http://www.sciencedaily.com/news/matter_energy/
solar_energy/

Solar Power World: http://www.solarpowerworldonline.com/

Solrico: http://www.solrico.com/en/downloads/articles.html

Aansen, Krister, Stefan Heck, and Dickon Pinner. 2012. "Solar Power: Darkest before Dawn." McKinsey & Company. http://www.mckinsey.com/client_service/sustainability/latest_thinking/solar_powers_next_shining. Accessed on March 31, 2015.

This report points out that some experts have been concerned about the drop-off of consumer demand for solar energy, but the company believes that this trend is an indication of growing pains in the industry rather than its "death throes." The publication contains an abundance of useful economic data with which to study this issue.

"About: SunShot." 2015. Office of Energy Efficiency and Renewable Energy. U.S. Department of Energy. http://energy.gov/eere/sunshot/about. Accessed on March 30, 2015.

This web page provides detailed information about the U.S. government's SunShot initiative, a program that seeks to make solar electricity competitive with other forms of electricity by the end of the 2010s.

Appleyard, David. 2009. "Action Plan for 50%: How Solar Thermal Can Supply Europe's Energy." Renewable Energy World.com. http://www.renewableenergyworld.com/rea/news/article/2009/04/action-plan-for-50-how-solar-thermal-can-supply-europes-energy. Accessed on March 31, 2015.

This article describes plans developed by the European Solar Thermal Technology Platform (ESTTP) to find ways of using solar thermal power to supply half of Europe's energy needs by the year 2050. The plan focuses on the development of active solar heating and cooling of buildings, the use of solar heat in industrial operations, district heating and cooling needs, and improved methods of distributing solar heating products.

Banoni, Vanessa, et al. 2012. "The Place of Solar Power: An Economic Analysis of Concentrated and Distributed Solar Power." *Chemistry Central Journal.* 6(Suppl. 1): 56. http://download.

springer.com/static/pdf/334/art%253A10.1186%252F1752-1
53X-6-S1-S6.pdf?auth66=1427754350_51e7a82bebf6d7b7a6
a9041e92d6de58&ext=.pdf. Accessed on March 30, 2015.

This article provides a detailed economic analysis of the
costs and benefits, both financial and environmental, of
two forms of solar energy: photovoltaic cells and Stirling
systems.

Barber, Scott. 2015. "History of Passive Solar Energy." http://
uncw.edu/csurf/Explorations/documents/ScottBarber.pdf.   Accessed on March 31, 2015.

This article provides an excellent and comprehensive review of the use of passive solar energy beginning in the
15th century BCE to the present day, with a useful general explanation of the methods used for the technology.

Blackburn, John O., and Sam Cunningham. 2010. "Solar and
Nuclear Costs—the Historic Crossover." NC WARN. http://
www.ncwarn.org/wp-content/uploads/2010/07/NCW-SolarRe
port_final1.pdf. Accessed on March 31, 2015.

The authors report on their economic analysis of the costs
of nuclear and solar power, which finds that 2010 marked
the year in which solar energy became less expensive than
nuclear energy as a way of producing electricity in the
United States. The article is at times somewhat technical for the general reader, but it is nonetheless extremely
valuable with regard to the conclusions it draws and the
implications it has for future policy decisions.

Chu, Jennifer. 2014. "Steam from the Sun." MIT News. http://
newsoffice.mit.edu/2014/new-spongelike-structure-converts-
solar-energy-into-steam-0721. Accessed on January 29, 2015.

A research team at the Massachusetts Institute of Technology (MIT) has invented a system that converts solar energy into steam. The system consists of layers of graphite
flakes and carbon foam, which, when soaked with water,

change the water into steam when bombarded with solar radiation. The amount of steam produced is proportional to the intensity of the solar radiation.

De Winter, Francis, Ronald B. Swenson, et al. 2006. "Dawn of the Solar Era." http://www.ecotopia.com/ases/solartoday/dawn ofthesolarera.pdf. Accessed on March 31, 2015.

This web page consists of four articles arguing that the era of fossil fuels is coming to an end and that solar energy is the best possible substitute for coal, oil, and natural gas. The four articles are "A Wake-up Call," "The Second Half of the Age of Oil Dawns," "Imagine," and "Transitioning to a New Paradigm."

Font, Vince. 2014. "The Solar Energy Outlook for 2014." Renewable Energy World.com. http://www.renewableenergy world.com/rea/news/article/2014/02/the-solar-energy-outlook-for-2014. Accessed on March 30, 2015.

This article reviews the current status of solar energy worldwide (as of 2014) and attempts to lay out probable directions for the technology in the short-term future.

Fthenakis, Vasilis, James E. Mason, and Ken Zweibel. 2008. "The Technical, Geographical, and Economic Feasibility for Solar Energy to Supply the Energy Needs of the US." Energy Policy. http://www.solarplan.org/Research/F-M-Z_Solar%20Grand%20Plan_Energy%20Policy_2009.pdf. Accessed on March 31, 2015.

The authors examine the scientific, technical, and economic potential of solar energy over the near future and find that the technology has the potential to supply 39 percent of the total electricity needs and 35 percent of the total energy needs of the United States by 2050 and more than 90 percent of the nation's total energy needs by 2100. This changeover would also result in a reduction of 92 percent of the energy-related carbon dioxide emissions compared to 2005 levels.

Garretson, Peter. 2012. "Solar Power in Space?" Strategic Studies Quarterly. http://www.au.af.mil/au/ssq/2012/spring/garretson .pdf. Accessed on March 31, 2015.

This article provides a detailed history and analysis of the possibility of using satellites for the capture of sunlight and its transmission to Earth as a source of energy. The author suggests that the technology is "either the most important space project of our generation—critical to securing American long-term interests and requiring the advocacy of Airmen—or a fool's errand, an impossible dream threatening to divert valuable resources from where they are most needed today."

"A Genuine 1870 Solar-Powered Steam Engine." 1975. Mother Earth News. http://www.motherearthnews.com/green-transportation/solar-powered-steam-engine-zmaz75ndzgoe.aspx. Accessed on February 6, 2015.

This article contains a reprint of an article that appeared in an 1870 issue of The Technologist magazine describing a solar-powered engine designed and built by Swedish-American inventor John Ericsson.

Gopinathan, C.K., and J.S. Sastry. 1992. "A Multi-Purpose Rural Development Programme for Coastal Regions Utilising Solar Energy and the Sea." *Physical Processes in the Indian Seas. Proc. 1 Convention ISPSO, 1990.* http://drs.nio .org/drs/bitstream/2264/3090/2/Proc_Phys_Process_Indian_ Sea_1992_219.pdf. Accessed on March 30, 2015.

This article presents a very interesting description of a project that uses solar energy to make use of seawater for a wide variety of application of applications in an oceanside community, including the production of freshwater, the collection of mineral salts, and the production of power for the community.

Grose, Thomas K. 2014. "Solar Chimneys Can Convert Hot Air to Energy, But Is Funding a Mirage?" National Geographic.

http://news.nationalgeographic.com/news/energy/2014/
04/140416-solar-updraft-towers-convert-hot-air-to-energy/. Accessed on March 31, 2015.

Solar chimneys are a device for converting solar energy into mechanical energy. Air near the ground is heated by sunlight and then allowed to flow upward through a tall chimney. As it passes up the chimney, it passes through turbines whose motion converts the energy of moving air into mechanical work that can be used to produce electricity. This article discusses the technology and the likelihood of its becoming economically feasible.

"Health and Safety Concerns of Photovoltaic Solar Panels." 2015. Oregon Department of Transportation. http://www.oregon.gov/odot/hwy/oipp/docs/life-cyclehealthandsafetyconcerns.pdf. Accessed on March 31, 2015.

This website begins with the statement that "[t]he generation of electricity from photovoltaic (PV) solar panels is safe and effective. Because PV systems do not burn fossil fuels they do not produce the toxic air or greenhouse gas emissions associated with conventional fossil fuel fired generation technologies. According to the U.S. Department of Energy, few power-generating technologies have as little environmental impact as photovoltaic solar panels." It goes on to say, however, that PV solar panels do have some *potential* safety hazards, which it goes on to discuss.

Hill, Joshua S. 2015. "SolarCity Launches GridLogic: A Global Microgrid Service." CleanTechnica. http://cleantechnica.com/2015/03/16/solarcity-launches-gridlogic-a-global-microgrid-service/. Accessed on March 31, 2015.

This article describes the new microgrid technology developed by SolarCity for the transmission of solar electricity to small and regional geographic areas.

Holm, Dieter. 2015. "Renewable Energy Future for the Developing World." International Solar Energy Society. http://

whitepaper.ises.org/ISES-WP-600DV.pdf. Accessed on March 31, 2015.

This white paper lays out a rational for the expanded use of renewable energy resources, such as solar energy, in developing nations. It focuses on existing policies that deal with these technologies and proposes other policies that will serve to increase the use of renewable energy sources in developing nations.

"How Solar Energy Works." 2015. Union of Concerned Scientists. http://www.ucsusa.org/clean_energy/our-energy-choices/renewable-energy/how-solar-energy-works.html#.VRl5EfnF-So. Accessed on March 30, 2015.

This website provides an excellent general introduction to the topic of solar energy, with some historical background and a good general introduction to various types of solar energy systems.

Hull, John Ralph. 1979. "Physics of the Solar Pond." Retrospective Theses and Dissertations. Iowa State University. http://lib.dr.iastate.edu/cgi/viewcontent.cgi?article=7607&context=rtd. Accessed on February 6, 2015.

This doctoral thesis provides a comprehensive introduction to the concept of solar ponds as a source of energy.

Jones, Geoffrey, and Loubna Bouamane. 2012. "'Power from Sunshine': A Business History of Solar Energy." Harvard Business School. http://www.hbs.edu/faculty/Publication%20Files/12-105.pdf. Accessed on February 6, 2015.

This paper provides a review of the development of solar energy since its earliest use in the mid-nineteenth century, with special attention to the business and economics aspects of the use of solar power.

Kind, Peter. 2013. "Disruptive Challenges: Financial Implications and Strategic Responses to a Changing Retail Electric Business." Edison Electric Institute. http://www.eei.org/ourissues/finance/documents/disruptivechallenges.pdf. Accessed on March 31, 2015.

This article deals with recent technological and economic changes (defined here as "disruptive challenges") to the electric utility industry, such as the increases use of solar-generated electrical power. The author attempts to describe the changes that are likely to occur in the financing, administration, and operation of the electric utility industry as a result of the new options posed by these "challenges."

Kniew, Gerhard. 2006. "Global Energy and Climate Security through Solar Power from Deserts." http://www.desertec.org/downloads/deserts_en.pdf. Accessed on March 31, 2015.

The author claims that solar energy available in the world's deserts is more than 700 times the amount used by the world in all other forms today, and that that energy could be captured and transmitted economically to 90 percent of the world's population.

Konttinen, Petri. 2015. "Last Word: Solar Thermal—Time to Redress the Balance." Renewable Energy World.com. http://www.renewableenergyworld.com/rea/news/article/2008/10/last-word-solar-thermal-time-to-redress-the-balance-53903. Accessed on March 31, 2015.

The author explores the question as to why solar thermal energy has not received the attention he thinks it deserves and points out the special advantages available from this form of solar technology.

La Roche, Pablo. 2006. "Lecture 5: Passive Heating." http://www.cpp.edu/~p3team/f-studio/lectures/P3%20PASSIVE%20Heating.pdf. Accessed on March 31, 2015.

This web article is one of a series of lectures in architecture by Professor La Roche. It provides a superb history of passive solar use in architectural design, with an excellent explanation of the principles involved in the technology.

Lenda, Chris. 2012. "Clarification on the Cost & Incentives of Solar Energy Systems." Connecticut Builder. http://www .aegis-solar.com/news/2012/CT%20Builder%20article.pdf. Accessed on March 31, 2015.

> This article provides an excellent overview of the technical aspects of various types of solar energy systems, written in a manner that can be easily understood by the general public.

Parlevliet, David, and Navid Reza Moheimani. 2014. "Efficient Conversion of Solar Energy to Biomass and Electricity." *Aquatic Biosystems*. 10: 4. (Online article) http://www.aquaticbiosystems. org/content/10/1/4. Accessed on January 29, 2015.

> This article proposes a method by which two systems for generating electricity from sunlight—photovoltaic cells and photosynthetic microalgae—can be used in concert to obtain a greater fraction of the solar energy available in sunlight.

Perlin, John. 2005. "Solar Evolution." http://www.californiaso-larcenter.org/history.html. Accessed on February 6, 2015.

> This excellent review of the history of solar energy is divided into three parts: photovoltaics, solar thermal, and solar architecture.

Robertson, Keith, and Andreas Athienitis. 2015. "Solar Energy for Buildings." https://www.cmhc-schl.gc.ca/en/inpr/bude/himu/co edar/upload/OAA_En_aug10.pdf. Accessed on March 31, 2015.

> This article provides detailed and technical information about incorporating solar energy factors into the design and construction of single unit and multiunit residential buildings, including passive solar heating, ventilation, air heating, solar domestic water heating, and shading.

Seltenrich, Nate. 2010. "Oakland Invades the Desert." East Bay Express. http://www.eastbayexpress.com/oakland/oakland-invades-the-desert/Content?oid=2258794. Accessed on February 2, 2015.

This article describes efforts to construct large solar power plants in the Nevada Desert, electricity from which will then be shipped to the East Bay region of northern California. The article outlines not only the benefits the system may provide for California cities but also the potential environmental problems such large industrial operations may pose for the regions in which they are being sited.

Sharma, Pragya, and Tirumalachetty Harinarayana. 2013. "Solar Energy Generation Potential along National Highways." *International Journal of Energy and Environmental Engineering.* 4(16). http://download.springer.com/static/pdf/627/art%253A10.118 6%252F2251-6832-4-16.pdf?auth66=1427751952_a0ecf1f3d 0875cb8d931a6e25a9f0fb5&ext=.pdf. Accessed on March 20, 2015.

The authors discuss the possibility of constructing roofs consisting of solar arrays above major highways as a way of making use of otherwise unused space for the production of electrical energy.

Skulls in the Stars. 2010. "Mythbusters Were Scooped—by 130 Years! (Archimedes Death Ray)." http://skullsinthestars .com/2010/02/07/mythbusters-were-scooped-by-130-years-ar chimedes-death-ray/. Accessed on February 5, 2015.

This fascinating posting reviews a century and a half worth of research as to the validity of the claim that Archimedes used a system of mirrors to destroy a fleet of Roman ships in the harbor of Syracuse in 212 BCE.

Smith, Charles. 2015. "History of Solar Energy." Solar Energy .com. http://solarenergy.com/power-panels/history-solar-energy. Accessed on February 6, 2015.

This article provides a readable and interesting account of the general features of the development of solar energy from the mid-nineteenth century to the present day.

"Solar." 2015. Energy Kids. U.S. Energy Information Administration. http://www.eia.gov/kids/energy.cfm?page=solar_home-ba sics. Accessed on March 30, 2015.

> This website is designed especially for younger children with basic information on solar energy, such as solar photovoltaics, solar thermal power plants, solar thermal collectors, and solar energy, and the environment.

"Solar." 2015. Office of Energy Efficiency and Renewable Energy. U.S. Department of Energy. http://energy.gov/science-in novation/energy-sources/renewable-energy/solar. Accessed on March 30, 2015.

> This webpage provides an array of basic information about solar energy, such as solar homes, the solar decathlon, very large solar systems, and the department's Sun-Shot initiative.

"Solar Energy." 2015. ATTRA Sustainable Agriculture. https:// attra.ncat.org/attra-pub/farm_energy/solar.html. Accessed on March 31, 2015.

> This web page provides links to a number of resources describing the application of solar energy to agriculture, such as solar greenhouses, solar-powered livestock watering systems, and food dehydration options.

"Solar Energy." 2015. Electric Light and Power. http://www.elp .com/topics/solar-energy.htm. Accessed on March 31, 2015.

> This website contains articles on current developments in solar electricity production, such as new facilities that are being planned or built and new laws and regulations being considered or adopted.

"Solar Energy." 1995. Energy Story. California Energy Commission. http://energyquest.ca.gov/story/chapter15.html. Accessed on March 30, 2015.

> This web page is chapter 15 of a book on energy published in 1995 by the California Energy Commission. It

provides basic information about solar photovoltaic, solar thermal electricity, and solar hot water in an easily understandable way for the general reader

"Solar Energy." 2015. Solar Energy.com. http://solarenergy.com/. Accessed on March 30, 2015.

This website claims to be "the authority of all things solar." It provides information on a wide variety of solar-related topics such as the benefits of solar panel systems for homes, how solar leasing works, state and federal tax incentives for the use of solar power systems, modern solar power technology, and solar electrical power events and expos.

"Solar Energy." 2015. Solar Energy.net. http://solarenergy.net/. Accessed on March 30, 2015.

This website is devoted to reports on recent developments in science, technology, development, economics, politics, and other aspects of solar energy. It provides an excellent update on current events in the field.

"Solar Energy." 2015. U.S. Department of the Interior. Bureau of Land Management. http://www.blm.gov/wo/st/en/prog/energy/solar_energy.html. Accessed on March 31, 2015.

This web page focuses on the construction and operation of solar facilities on land owned by the federal government and administered by the Bureau of Land Management. It provides technical information on development policy, effects on wildlife, federal funding and incentives, right-of-way issues, and related topics.

"Solar Energy 101." 2015. Dow Corning. http://www.dowcorning.com/content/solar/solarworld/solar101.aspx. Accessed on March 31, 2015.

This web page (and its companion website, "Solar Energy 201") provides an easily understood and comprehensive explanation of solar photovoltaic systems and their applications in residential homes.

"Solar Energy Basics." 2015. National Renewable Energy Laboratory. http://www.nrel.gov/learning/re_solar.html. Accessed on March 30, 2015.

This NREL site provides a complete and clearly written description of the major types of solar energy: solar photovoltaic, concentrating solar power, solar process heat, passive solar technology, and solar water heating. The page also provides links to other useful sites.

"Solar Energy News." 2015. Science Daily. http://www.sciencedaily.com/news/matter_energy/solar_energy/. March 30, 2015.

Science Daily is one of the best online sources of current information on a wide variety of scientific topics. This web page focuses on recent developments in solar energy.

"Solar Water Heating Basics." 2015. Home Power. http://www.homepower.com/articles/solar-water-heating/basics/what-solar-water-heating. Accessed on March 31, 2015.

This article explains what solar water heating is, the advantages of the technology, how solar water-heating systems operate, and how they are installed.

"Up with the Sun." 1996. Union of Concerned Scientists. http://www.ucsusa.org/clean_energy/smart-energy-solutions/increase-renewables/up-with-the-sun-solar-energy.html#.VRmKcPnF-So. Accessed on March 30, 2015.

This web page reviews a variety of ways in which solar energy can be used in agricultural operations. The information is taken from the UCS fact sheet on the same topic, which can be accessed from the web page.

"Using Solar Energy for Air Conditioning and Heating." 2015. AC Southeast. http://airconditioningsoutheast.com/info/article/

using-solar-energy-for-air-conditioning-and-heating. Accessed on March 31, 2015.

This excellent web page provides detailed information on the nature of solar energy systems and the way they can be used in both air conditioning and heating systems for homes, offices, and other structures.

Yirka, Bob. 2011. "India Signs on to Floating Solar Energy Power Plant (w/video)." Phys.Org. http://phys.org/news/2011-03-india-solar-energy-power-video.html. Accessed on March 31, 2015.

This article discusses plans for the development of floating solar photovoltaic systems in India, a technology that is being proposed for a number of locations where sufficient land areas are generally not available for the construction of such projects.

7   Chronology

Introduction

Humans have been aware of the possibilities for using energy from the Sun for at least 8,000 years. Over that period of time, they have developed a variety of technologies for using that energy to heat homes and commercial buildings, drive agricultural and industrial processes, generate electricity, and carry out a number of other essential procedures in human life. The steps forward in this long history have depended on a number of crucial discoveries in basic science and technology, which have made solar energy the promising part of the world's economies today. This chapter summarizes some of the important events that have occurred in the history of solar energy.

15th century BCE   Egyptian pharaoh Amenhotep III is said to have had constructed two "sounding statues," now known as the Colossi of Memnon, that produced sounds when air that had been heated by sunlight escaped through holes in their base. At the same time, solar energy was apparently being used for heating homes, distilling water, and drying agricultural products.

ca. 7th century BCE   Plutarch (45–120 CE) explains that sacred fires used during this period are sometimes extinguished by wars or other events, and must then be rekindled *only* by

---

A modern building in California utilizes solar energy cells that are incorporated into the architecture. (Paul Matthew Photography/Shutterstock.com)

303

"lighting a pure and unpolluted flame from the rays of the sun" (*Numa*, Chapter 9, section 6).

7th century BCE   The town of Olynthus is constructed in the region of Chalcidice in northeastern Greece. It is said to be the first solar city because all streets and houses in the city were aligned so as to permit the sun's rays to enter homes during the winter, providing them with heat, and to prevent sunlight from entering during the summer, keeping them cool during that part of the year.

424 BCE   The use of a magnifying glass to focus the Sun's rays is reputedly first mentioned in this year in a play by Aristophanes, *The Cloud*, where such a glass was said to be for sale at a pharmacy.

ca. 3rd century BCE   Historians differ about the exact dates involved, but evidence suggests that people in many parts of the world were beginning to use devices for the concentration of solar radiation to light torches for the lighting of dwellings in the evening as well as for religious and other ceremonial purposes.

ca. 270 BCE   The Greek architect Sostratus of Cnidus designs the Great Lighthouse of Alexandria, one of the Seven Wonders of the Ancient World. At the top of the 384-foot tower that makes up the lighthouse is a system of mirrors that reflects sunlight during the day, reputedly to a distance of up to 50 miles, and the light of burning oil at night.

ca. 212 BCE   Greek inventor and mathematician Archimedes is credited with inventing a system of mirrors that focused on and concentrated the Sun's rays on a fleet of Roman ships in the harbor of Syracuse. The heat produced by the mirrors was reputedly sufficient to set fire to the ships, breaking the siege of the city. (The validity of this claim has been the subject of a seemingly endless critique by professional and amateur scientists ever since at least 1867.)

89 BCE   After Rome takes control of the city of Herculaneum, some parts of the city are rebuilt so that homes always face toward the south to take advantage of solar energy to heat the houses.

ca. 30 BCE   The Emperor Tiberius orders the construction of a protected structure in which cucumbers can be grown year-around for (depending on the source) the maintenance of his good health or his pure enjoyment of the vegetable. The structure is the earliest known greenhouse, called, in the Latin, a *specularium*. It is heated by solar radiation captured thin sheets of mica, sheet glass not yet having been invented.

1st century CE   Roman natural philosopher and author Pliny the Younger orders a room to be added to his home that takes advantage of solar radiation to produce continuous heat, a room he calls his *heliocaminus*, or "solar furnace."

529–534   The Justinian Code includes a section prohibiting anyone from building a structure or allowing trees to grow in such a way as to block out the sunlight needed to heat an adjacent solar passive home (heliocaminus).

1615   French physicist and engineer Salomon de Caux is credited by some historians as constructing the first solar device since classical times, a water pump powered by solar radiation that heats water in an enclosed container, causing the water to boil and shoot out of an opening in the container.

1767   Swiss physicist Horace Benedict de Saussure is credited with inventing the first solar oven, a device for concentrating the Sun's rays to a point where sufficient heat is produced to cook foods.

1839   Nineteen-year-old French physicist Edmund Becquerel discovers the photoelectric effect when, working with a simple electrolytic cell, he observes that the flow of electrons produced in the cell is increased when it is exposed to a beam of sunlight. He also coins the term *photoelectric effect* for the phenomenon he has observed.

1866   French inventor Augustin Mouchot invents the first parabolic trough solar collector, one of the primary devices used in concentrating solar power (CSP) systems today. Convinced that fossil fuels would soon be depleted and unavailable as a source of energy, Mouchot devoted his life to inventing a variety of solar devices to replace coal, a body of work summarized in his book *La Chaleur solaire et ses Applications industrielles (Solar Heat and Its Industrial Applications)* in 1869.

1870   Swedish-American inventor and engineer John Ericsson invents an early solar-powered engine that uses a parabolic-shaped collector to use sunlight to boil water that drives the machine. The device is designed to move water through irrigation systems in the largely undeveloped western United States.

1872   The first commercial solar-powered water distillation plant is designed by American engineer Charles Wilson and constructed in Las Salinas, Chile. The plant produced 6,000 gallons of pure water per day that was used for watering mules who worked in adjacent mining operations.

1873   British telegraphic engineer Willoughby Smith discovers that the element selenium is photoconductive. In looking for materials with a relatively high resistance to the flow of electrical current, but that were not also complete insulators, Smith found that selenium conducts only a modest electrical current in the dark, but a greatly increased flow in the presence of sunlight.

1883   American inventor Charles Fritts is credited with inventing the first selenium-based photovoltaic cell. Although he hoped his cell would compete with Thomas Edison's coal-fired power plants in the production of electricity, his cells had an efficiency of only about 1 percent and thus were not an economical source of electrical power.

1887   During his experimental studies on James Clerk Maxwell's theory of electromagnetism, German physicist Heinrich Hertz observes that shining ultraviolet light on a metal ball results in an increase in the flow of electrical current from the ball, the first recorded observation of the photoelectric effect.

Hertz was not particularly interested in the phenomenon, and conducted no further research on it (see also 1902).

1887 German physicist James Moser discovers that the photoelectric effect is enhanced by the addition of erythrosine dye to a solar cell, a discovery that forms the basis of modern dye-sensitized photoelectrochemical cell systems.

1891 Baltimore resident Clarence Kemp, seller of home heating equipment, invents the first commercial solar water heater, combining earlier "hot box" systems for concentrating solar energy to heat a space and metal tanks painted black to absorb sunlight for producing hot water.

1892 American businessman Aubrey Eneas founds the Solar Motor Company, for the purpose of building engines that operate on solar power rather than the burning of coal. He expected to sell his solar motors in the American West, where coal was not plentiful, for the purpose of moving irrigation water. Unpredictable weather patterns made the project uneconomical, however, and the company went out of business in 1905.

1902 American inventors and entrepreneurs John Boyle and H.E. Willsie construct the first experimental solar power plant in Olney, Illinois. The plant makes use of a shallow wooden tank covered with a double layer of window glass, surrounded by enclosed air spaces filled with hay. The system was able to function efficiently even on cloudy days but was unable to compete economically with existing coal-fired power plants.

1902 German physicist Philipp Lenard follows up on Hertz's discovery of the photoelectric effect by testing the result of shining light of different colors and intensities on the surface of a metallic plate, earning him the title of father of the photoelectric effect. Lenard learns that the velocity of electrons leaving the metal surface is a result of the wavelength of the incident light, not its intensity.

1903 Spanish military engineer Isidoro Cabanyes suggests the construction of a solar updraft tower, in which large masses of air are heated in a greenhouse-type setting at the base of a tall tower, inducing convection currents that escape upward through the tower. At the top of the tower, a fan is caused to rotate by the escaping air, inducing an electric current in an attached generator.

1904 Willsie constructs the first solar power storage system at two of his solar power plants, in St. Louis and Needles, California. Water that is heated by sunlight during the day is kept warm at night in insulated ponds and is then used to continue operating a sulfur-dioxide pump system, throughout evening hours.

1905 Austrian American physicist Albert Einstein provides a theoretical explanation for the photoelectric effect, an accomplishment for which he receives the 1921 Nobel Prize in Physics.

1907 American inventor Frank Shuman attempts to improve on existing solar power steam engines by replacing water with ether as the working fluid. Although ether has a much lower boiling point than water, its specific gravity is too low to serve as a working fluid, and Shuman soon returns to the use of water as a working fluid. He improves the efficiency of his engines, instead, by using better insulation systems to retain heat within the engine.

1918 Polish chemist and metallurgist Jan Czochralski accidentally discovers a method for growing large, single crystals of a metal, a procedure that is later adapted to the growing of large, single crystals of silicon for use in the manufacture of solar devices.

1943 American architect George Lof designs the first completely solar-powered home in the United States, a structure built in Boulder, Colorado.

1946 American architect Arthur Brown designs the first completely solar-powered public building in the United States, the Rose Elementary School in Tucson, Arizona.

1946 Inventor Russell S. Ohl receives a patent for the first solar cell, a device that converts visible light into an electric current. (The patent is assigned to Bell Laboratories.)

1954   Bell Laboratories researchers Daryl Chapin, Calvin Fuller, and Gerald Pearson invent the first practical photo-voltaic solar cell (solar battery) for converting sunlight into useful electrical power. The efficiency of the battery is only about 6 percent, but that is sufficient for the manufacture of the first solar panels later used on virtually all artificial satellites.

1955   General Motors engineer William Cobb designs what is thought to be the first electric vehicle, a model car 15 inches in length.

1956   American architect Frank Bridgers designs the first commercial solar building, the Solar Building, in Albuquerque, New Mexico. The building is converted into conventional fossil-fuel energy when it is expanded in 1962.

1958   Israeli researcher Rudolph Bloch suggests the possibility of building artificial non-convecting solar salt ponds for the production of thermal energy from solar radiation. Follow-up research on this suggestion was later carried out by Israeli physicist Harry Zvi Tabor, who produced the first such pond a year later on the shores of the Dead Sea.

1958   Solar cells produced by the Hoffman Electronics company reach a record efficiency of 9 percent.

1958   The space satellite *Vanguard I* is launched, powered by both chemical batteries (which last only a little more than two weeks) and solar batteries (which last more than six years).

1960   Hoffman Electronics solar cells reach an efficiency of 14 percent.

1968   Inventor Roger Riehl invents what is thought to be the first solar-powered wrist watch, which he calls the Synchronar.

1973   The Arab Organization of Petroleum Exporting Countries (OPEC) initiates an oil embargo on nations that have supported Israel in its conflict with Arab nations, creating a new impetus in the search for alternative forms of energy in nations affected by the embargo.

1973   President Richard M. Nixon announces Project Independence 1980, a program designed to make the United States energy-independent by that year. The program never came close to achieving that goal.

1974   The first solar-powered aircraft, the 27-pound AstroFlight Sunrise I, is launched from Fort Irwin Military Reservation.

1974   The first state tax incentives for the installation of solar energy facilities are adopted in Indiana and Arizona.

1974   The U.S. Congress passes the Energy Reorganization Act of 1974, which, among other provisions, establishes the Energy Research and Development Administration (ERDA). Among its responsibilities is responsibility for all federal renewable energy projects.

1974   The U.S. Congress passes the Solar Energy Research, Development and Demonstration Act of 1974, which, among other provisions, creates the Solar Energy Research Institute (SERI), later to become the National Renewable Energy Laboratory in 1991.

1974   The U.S. Congress passes the Solar Heating and Cooling Demonstration Act of 1974.

1978   The state of California creates a commission called SolarCal to encourage the use of solar energy in the state. An important component of the new plan is a 55 percent investment tax credit for the purchase of solar energy equipment.

1978   The U.S. Congress passes the Energy Tax Act of 1978, which, among other sections, provides for tax credits for the installation of solar devices and other forms of renewable energy implementation.

1979   In at least a partially symbolic move in support of solar energy, President Jimmy Carter orders that solar panels be installed on the White House roof.

1979   The first manned solar-powered spacecraft, *Mauro Solar Riser*, is launched at Flabob Airport in Riverside, California.

1980 The election of President Ronald Reagan marks the beginning of a new era in which the role of renewable energy resources, including solar power, is dramatically diminished in comparison with trends of the preceding decade.

1991 The Solar Energy Research Institute is renamed and its mission broadened in the formation of the National Renewable Energy Laboratory (NREL).

1992 The Energy Policy Act of 1992 creates a category of tax incentive known as production tax credits (PTC) that can be used by individuals and companies that produce certain types of renewable energy. Solar energy is not added to the program until the adoption of the American Jobs Creation Act of 2004.

2004 Solar energy is added to the production tax credit program for the first time (see also **1992** and **2005**).

2005 Solar energy is removed from the production tax credit program as part of the Energy Policy Act of 2005.

2006 The Solar Aid charity creates a program called Sunny Money for selling solar-powered lanterns in Africa, where nearly 600 million people do not have access to electricity. The purpose of the project is to help people replace traditional kerosene lanterns with solar-powered devices.

2010 The U.S. Department of Energy launches the SunShot program, a program designed to sponsor research, development, education, and related projects that will reduce the cost of solar electricity by 75 percent compared to 2010 prices by the year 2020.

2010 President Barack Obama orders that solar panels once more be installed on the roof of the White House.

2011 Google announces that it will contribute $168 million toward the construction of a new large solar facilities in the Mojave Desert of California. The greatest portion of the plant's cost of $2.2 billion comes from a $1.6 billion loan from the federal government.

2011 The U.S. solar cell manufacturing company Solyndra goes bankrupt, creating a controversy about the wisdom of providing government-backed funding for solar industries. Later research shows that, overall, such programs actually return a profit to the federal government.

2014 India announces a new initiative to expand the nation's solar energy capacity by 30 times by the year 2020.

2014 The world's largest solar power plant begins operation in San Luis Obispo County, California. With a rated capacity of 550 megawatts, the plant is capable of meeting all electrical needs of at least 110,000, and perhaps as many as 176,000, homes.

2015 Japanese researchers report success in the design and implementation of technology used for the wireless transmission of solar energy from space satellites to Earth-based receiving stations.

2015 The solar-powered aircraft *Solar Impulse 2* begins an around-the-world trip from Abu Dhabi carrying one pilot in a time. The flight reaches Hawaii before being delayed by damage to its batteries. Plans call for it to complete its trip in early 2016.

Introduction

Understanding the history, background, technology, and issues related to solar energy depends to a substantial degree in having command of the vocabulary used in discussing the topic. This chapter offers definitions for some of the most commonly used terms in the field of solar energy. A number of good glossaries are also available on the Internet, such as those provided by the U.S. Department of Energy's Office of Energy Efficiency & Renewable Energy (http://energy.gov/eere/sunshot/solar-energy-glossary), the U.S. Energy Information Administration (http://energy.gov/eere/sunshot/solar-energy-glossary), and PV Power.com (http://www.pvpower.com/glossary.html).

Alternative energy resource   Any form of energy other than a fossil fuel. The term is usually used for all forms of renewable energy resources (*q.v.*) as well as nuclear resources.

Amorphous silicon   A form of the element silicon that has no crystalline structure. Often designated as a-Si.

Array (photovoltaic)   A collection of photovoltaic panels and/or modules that functions as a single unit to convert sunlight into electrical current.

Concentrating solar power (CSP)   Any system that uses mirrors to reflect and concentrate sunlight for the purpose of producing heat, which, in turn, is used to produce electricity.

Crystalline silicon   A form of the element silicon that has crystalline structure. Often designated as c-Si.

CSP   *See* Concentrating solar power.

Czochralski process   A method of growing large, high-quality semiconductor crystals by slowly lifting a seed crystal from a molten bath of the material under careful cooling conditions. Also known as Czochralski method.

Distributed energy resources (DERs)   Power sources of relatively small size that can be brought together to provide electrical power needed to meet regular demands.

Dopant   A substance added to an otherwise pure semiconducting material in order to change the electrical properties of that material.

Electric grid   *See* Grid.

Energy   In science, the ability to do work, that is, the ability to move an object over a distance. *See also* Power.

Energy equation   A term sometimes used to describe all the energy inputs and outputs that take place in the world, in specific nations or states, or in other geographic or political entities.

Fixed tile array   A photovoltaic system that is set at a fixed angle to the horizontal.

Flat plate array   A photovoltaic system in which the modules are arranged in a fixed, flat conformation. Such a system is incapable of collecting the maximum amount of solar radiation at all times of the day.

Fresnel lens   An optical device consisting of concentric rings that focus light at a single point.

Generator nameplate capacity   *See* Nameplate capacity.

Gigawatt (GW)   A unit of power equal to one billion watts.

Grid (power)   A set of transmission lines that carry electrical power from a production site, such as a coal-fired power plant, to consumers. The grid also includes the stations needed to control the flow of electricity through the system.

Heliostat   A device that contains a mirror that can be rotated or otherwise moved in order to reflect sunlight to a collecting device.

Hybrid system (solar)   A system of energy sources that is combined to produce the maximum possible amount of energy over a variety of conditions, such as solar collectors combined with wind turbines.

Insolation   The total amount of solar radiation received on a given area of Earth's surface over a given period of time. The usual units of insolation are megajoule per square meter (MJ/m²), joule per square millimeter (J/mm²), or watt-hour per square meter (Wh/m²).

Installed capacity   *See* Nameplate capacity.

Intermittent energy source   *See* Renewable energy source.

Irradiance   A measure of insolation per unit of time. Usually expressed as watts per square meter (W/m²).

Kilowatt (kW)   A unit of power equal to one thousand watts.

Langley (L)   The unit of solar irradiance, equal to one gram calorie per square centimeter (85.93 kWh/m²).

Levelized cost of electricity (LCOE)   The cost of electricity produced by a solar system based on the system's installed price, its total lifetime cost, and its lifetime electricity production.

Megawatt (MW)   A unit of power equal to one million watts.

Module (photovoltaic)   A system consisting of one or more photovoltaic cells and their connections that is, itself, the basic unit of a photovoltaic panel (*q.v.*).

Nameplate capacity   The maximum electrical output of a generator as determined by the manufacturer of the equipment, generally expressed in megawatts (MW). Also called *generator nameplate capacity*, *installed capacity*, *nominal capacity*, or *rated capacity*.

Net metering   A billing mechanism in which a utility credits consumers for electricity they add to the utility's grid from their own private collecting system.

Nominal capacity   *See* Nameplate capacity.

Panel (photovoltaic)   A collection of photovoltaic modules that is often itself a part of a larger system known as a photovoltaic array.

Photovoltaic   A noun or adjective referring to the process by which solar energy is converted into electrical energy.

Polycrystalline silicon   A sample of the element silicon that consists of many individual silicon crystals. By contrast, *see also* Single-crystal silicon.

Power   In science, the rate at which work is done. The usual units of power are work done per second or work done per hour.

Pumped hydro   A system for storing electrical energy by using electricity produced from one source, such as solar energy, to pump water to a higher elevation (thereby adding to its potential energy), which can then be allowed to flow back downhill again to produce energy (in the form of kinetic energy) during times when the primary energy source is not available.

Rated capacity   *See* Nameplate capacity.

Renewable energy resource   Any form of energy, such as wind or solar, that is not depleted as it is used. *See also* Alternative energy resource.

Single-crystal silicon   A form of the element silicon in which the sample consists of a single crystal. Also known as *monocrystalline silicon*. By contrast, *see* Polycrystalline silicon.

Solar inverter   A device for converting direct current (DC) electricity produced by a solar collecting device into alternating current (AC) electricity that can be used in a utility grid.

Thin film   A layer of semiconductor material a few microns or less in thickness, used to make certain kinds of photovoltaic cells.

Variable generation source   Any source of energy that changes over time as a result of factors beyond the control of the system itself. Wind and solar energy are examples of variable-generation sources because neither the wind nor the sun is available dependably at all times of the day and night, year-around. Also known as *intermittent energy source*.

Watt (W)   A unit of power equal to the transport of one ampere of electric current at a potential different of one volt.

Index

About the Author

David E. Newton holds an associate's degree in science from Grand Rapids (Michigan) Junior College, a BA in chemistry (with high distinction), an MA in education from the University of Michigan, and an EdD in science education from Harvard University. He is the author of more than 400 textbooks, encyclopedias, resource books, research manuals, laboratory manuals, trade books, and other educational materials. He taught mathematics, chemistry, and physical science in Grand Rapids, Michigan, for 13 years; was professor of chemistry and physics at Salem State College in Massachusetts for 15 years; and was adjunct professor in the College of Professional Studies at the University of San Francisco for 10 years.

The author's previous books for ABC-CLIO include *Global Warming* (1993), *Gay and Lesbian Rights: A Resource Handbook* (1994, 2009), *The Ozone Dilemma* (1995), *Violence and the Mass Media* (1996), *Environmental Justice* (1996, 2009), *Encyclopedia of Cryptology* (1997), *Social Issues in Science and Technology: An Encyclopedia* (1999), *DNA Technology* (2009), *Sexual Health* (2010), *The Animal Experimentation Debate* (2013), *Marijuana* (2013), *World Energy Crisis* (2013), *Steroids and Doping in Sports* (2014), *GMO Food* (2014), *Science and Political Controversy* (2014), *Wind Energy* (2015), and *Fracking* (2015). His other recent books include *Physics: Oryx Frontiers of Science Series* (2000), *Sick!* (4 volumes) (2000), *Science, Technology, and Society: The Impact of Science in the 19th Century* (2 volumes; 2001), *Encyclopedia of Fire* (2002), *Molecular*

Nanotechnology: Oryx Frontiers of Science Series (2002), *Encyclopedia of Water* (2003), *Encyclopedia of Air* (2004), *The New Chemistry* (6 volumes; 2007), *Nuclear Power* (2005), *Stem Cell Research* (2006), *Latinos in the Sciences, Math, and Professions* (2007), and *DNA Evidence and Forensic Science* (2008). He has also been an updating and consulting editor on a number of books and reference works, including *Chemical Compounds* (2005), *Chemical Elements* (2006), *Encyclopedia of Endangered Species* (2006), *World of Mathematics* (2006), *World of Chemistry* (2006), *World of Health* (2006), *UXL Encyclopedia of Science* (2007), *Alternative Medicine* (2008), *Grzimek's Animal Life Encyclopedia* (2009), *Community Health* (2009), *Genetic Medicine* (2009), *The Gale Encyclopedia of Medicine* (2010–2011), *The Gale Encyclopedia of Alternative Medicine* (2013), *Discoveries in Modern Science: Exploration, Invention, and Technology* (2013–2014), and *Science in Context* (2013–2014).